CARTESIAN SPACETIME

ARCHIVES INTERNATIONALES D'HISTOIRE DES IDÉES

INTERNATIONAL ARCHIVES OF THE HISTORY OF IDEAS

181

CARTESIAN SPACETIME

by

EDWARD SLOWIK

CARTESIAN SPACETIME

Descartes' Physics and the Relational Theory of Space and Motion

by

EDWARD SLOWIK
*Winona State University,
Winona, U.S.A.*

KLUWER ACADEMIC PUBLISHERS
DORDRECHT / BOSTON / LONDON

A C.I.P. Catalogue record for this book is available from the Library of Congress.

ISBN 1-4020-0265-3

Published by Kluwer Academic Publishers,
P.O. Box 17, 3300 AA Dordrecht, The Netherlands.

Sold and distributed in North, Central and South America
by Kluwer Academic Publishers,
101 Philip Drive, Norwell, MA 02061, U.S.A.

In all other countries, sold and distributed
by Kluwer Academic Publishers,
P.O. Box 322, 3300 AH Dordrecht, The Netherlands.

Printed on acid-free paper

Printed in the Netherlands.

To my two most loyal and dedicated supporters,

EMILY SLOWIK

And

CHRISTINE KING

TABLE OF CONTENTS

PREFACE

This Project, long in gestation, owes a great deal to the help and support of numerous friends and colleagues over the past several years. Among the many whom I would like to thank for their comments and discussions: Ty Lightner, Will Alabran, George Pappas, Robert Batterman, John Carriero, Roger Florka, Helen Hattab, Paul Hoffman, Marleen Rozemond, Alan Nelson, Margaret Atherton, Andrew Black, and Dennis Des Chene. In particular, I owe a great deal to the invaluable assistance and friendship of Ronald Laymon, Jonathan Bennett, and Daniel Garber: without their help, the philosophical content of this work would have been significantly diminished (although I take full responsibility for the final published results). Above all, however, I have two people to thank for their friendship and professional guidance: Calvin Normore and Mark Wilson. Their vast store of knowledge, insights, and ideas in the history and philosophy of science served as the catalyst for this project, and I am greatly indebted and thankful for the opportunity to have worked so closely with them over the years. Moreover, although many portions of this work were directly inspired by Prof. Wilson's previous research, it should be noted that the conclusions reached in this investigation often differ quite considerably from his own views and conclusions (as is most evident in chapter 9).

Previous portions of the work have appeared, in an earlier form, in several philosophical journals. I grateful to the following journals for granting me permission to use this material: the appendix to chapter 3 appeared as, "Descartes and Circular Inertia," in *The Modern Schoolman,* 77, 1999, 1-11; part of chapter 4 appeared in, "Perfect Solidity: Natural Laws and the Problem of Matter in Descartes' Universe," *History of Philosophy Quarterly,* vol. 13, no. 2, April 1996, 187-204, and, "Descartes and Individual Corporeal Substance," in *British Journal for the History of Philosophy,* vol. 9, no. 1, 1-17; sections of chapter 5 in, "Descartes' Quantity of Motion: 'New Age' Holism meets the Cartesian Conservation Principle," in *Pacific Philosophical Quarterly;* vol. 99, no. 2, June 1999, 178-203; portions of chapter 6 in, "Descartes, Space-Time, and Relational Motion," in *Philosophy of Science,* 66 (March 1999) 117-139; portions of chapter 7 in, "Descartes' Forgotten Hypotheses on Motion: Kinematic Logic and

Relational Transfer," forthcoming in *Journal of Philosophical Research;* portions of chapter 8 in, "Huygens' Center-of-Mass Space-Time Reference Frame: Constructing a Cartesian Dynamics in the Wake of Newton's '*De Gravitatione*' Argument," in *Synthese,* 112: 247-269, 1997; and finally, portions of chapter 9 in, "Cartesianism and the Kinematics of Mechanisms: Or, How To Find Fixed Reference Frames in a Cartesian Space-Time," in *Noûs* 32:3 (1998) 364-385. Unless otherwise noted, all of the figures (illustrations) come from the Adam and Tannery complete edition of Descartes' *Oeuvres,* with each illustration linked to the specific citation provided in the text. The original illustrations/figures are not marked, however (namely, figures 1-7, 10, 17, 21, 24, 26-30).

INTRODUCTION

Although Descartes' natural philosophy marked an important advance in the development of modern science, many of his specific concepts of science—most notably, his concepts of space, time, and motion—have been largely discarded, and consequently neglected, since their introduction in the seventeenth century. Many critics over the years, such as Newton (in his early paper *De gravitatione*), have presented a series of apparently devastating arguments against Descartes' theory of space and motion; a generally negative historical verdict which, moreover, most contemporary scholars seem happy to accept. Nevertheless, it is also true that most historians and philosophers of science have not sufficiently investigated the viability and effectiveness of the case against Cartesian space and motion. This book, consequently, explores the success of the arguments against Descartes' theory of space and motion by determining if it is possible to formulate a Cartesian theory of science that can avoid the problems raised by Newton, as well as by a host of other past and contemporary philosophers either sympathetic or hostile to the overall design of Cartesian science. In particular, if Descartes' theory of motion can be reconciled with his hypotheses on dynamics (i.e., the branch of physics that deals with the motions of bodies under forces), then a Cartesian can successfully resolve the difficulties imposed by his critics. This book will comprise, therefore, a sustained attempt to construct a consistent "Cartesian" spacetime theory: i.e., a theory of space and time that incorporates Descartes' various theories of physics and nature without falling into the kinds of problems and contradictions as charged by, for example, Newton. Amazingly, the present investigation comprises the only known (book-length) treatise that exclusively focuses upon Descartes' theory of space and motion, as well as the possibility of its rehabilitation in the wake of much later criticism. Given the extremely important contribution of Descartes' theory of space, time, and motion to later such theories (such as Newton's, ironically), and given Descartes' preeminent importance in the overall history of Western philosophy, it is incredible that no previous commentators have devoted an exhaustive study of the potential for resurrecting Descartes' concept of space, time, and motion.

More specifically, the central difficulties for the devoted Cartesian are nicely encapsulated in Newton's argument (in his early paper, *De gravitatione*) that a theory of space and time can only meaningfully explicate the motions of bodies if it is equipped with fixed spatial locations. When endowed with such structures, a body's motion can be ascertained by simply determining the number of spatial positions traversed by the body during the temporal period spanned by its motion. Only when spatial positions are construed in this "absolute" sense—that is, as independently existing entities over and above the existence of objects—can motion be coherently described. However, Descartes held that space and time were purely "relational," an hypothesis that consigns the existence of space and time to mere relations among bodies (thus, they are nothing over and above these material relations). Overall, absolute spatial positions violate the tenets of a relational theory, since they regard such locations as fixed independently of the relative positions and motions of bodies. Hence, Newton concludes that Descartes cannot consistently employ the concept of motion in his relationalist scientific treatises, of which the most important of these scientific works is the *Principles of Philosophy*.

In Part I, after providing the details of Newton and Descartes' views on space, time, and motion, we shall investigate the essential elements of the absolute and relational conceptions of space and time. Rather than posit absolute spatial positions or absolute velocities, it will be shown that an absolute, or substantivalist, theory of space need only accept an absolute notion of acceleration. Modern formulations of Newton's theory will reveal that his conception of space is essentially "too strong" or overly rigid, and that a workable substantivalist hypothesis (such as Neo-Newtonianism) can effectively describe the motions of bodies with a much leaner ontology. In this context, moreover, I will advance one of the principal conclusions of this book; namely, that Newton's argument presupposes that a coherent theory of space and time must possess the capacity to compare information on bodily states across time. More precisely, if a spacetime theory desires to explicate the phenomena of bodily motion (velocity, acceleration, etc.), it must be equipped with the necessary "structure" to make meaningful comparisons of the dispositions and displacements of bodies at different temporal instants. This deeper, implicit element of Newton's argument I will deem a "background geometrical structure." This investigation of Newton's argument is necessary for our critique of Descartes, moreover, since it will be argued that there exist potential formulations of a Cartesian spacetime that actually meet Newton's minimal requirement for a "geometrical background structure". That is, *when properly construed,* Newton's claims (as well as the criticisms of many other philosophers) do not necessarily pose a threat to Descartes' overall conception of space and motion. In short, it may be possible for a Cartesian to accept Newton's limited demand for a geometrical structure *without* undermining Descartes' theory of space and motion.

As for the possibility of a Cartesian dynamics that can avoid the inconsistencies that appear intrinsic to Descartes' approach (as revealed by Newton, for example), this book will investigate a plethora of diverse formulations, mainly in Part III. The primary goal of these reconstructions is to harmonize Descartes' relational theory of space and time with (1) his views on the interactions of bodies, especially (although not necessarily) as exemplified in his seven collision rules, and (2) his conservation law, which holds that the quantity of motion (size times speed) of all bodies in the universe is conserved.

While Part II will largely concern important historical issues in the interpretation of Descartes' scientific concepts (as will be discussed at greater length below), chapter 6 will continue the exploration of the possibility of constructing a Cartesian spacetime (that we began in chapter

2). In particular, chapter 6 will strive to answer the question: If Descartes' theory does constitute some form of relationalism with regards to space and motion, then which type of relationalism is it? As recent work in the philosophy of space and time has revealed (e.g., John Earman), there are several discernible variations on a relationalist theme, thus it seems legitimate to inquire into the precise details of the *type* of relational theory of motion that Descartes' allegedly employed. As in chapter 2, our investigation will explore the problem of Cartesian motion in the light of contemporary work on the structure of theories of space, time, and motion—an analysis which, to the best of my knowledge, has largely neglected Descartes in favor of his later rival, Newton. Consequently, despite any historian's misgivings, it will be assumed that modern conceptual tools, if exercised judiciously, can be quite effective in revealing hidden facets of such historical theories. In fact, chapter 6 will demonstrate that Descartes' theory of motion does not exactly fit the strict relationalism that many commentators have traditionally believed; a conclusion that some contemporary scholars, such as D. Garber and D. Des Chene, have recently pointed out, as well. Nevertheless, our modern approach to the problem of Cartesian motion will also reveal the shortcomings of many previously attempted solutions, including both Garber's and Des Chene's. In the final sections of chapter 6, a deeper, underlying reason for the limited success of most earlier readings of Descartes' theory of space and motion will be disclosed: namely, commentators have neglected the dynamics of Descartes' system (which pertains to bodily motions under the actions of forces) in favor of the kinematics (which concerns bodily motion per se). More specifically, whereas most commentators have attempted to treat the kinematical problems of Cartesian physics before dealing with its numerous dynamical inconsistencies, a more serviceable solution to the problem of Descartes' kinematics might take the dynamics as primary, and thus reverse the course of the standard interpretation. As will be demonstrated, the (dynamic) natural laws, and the Cartesian collision rules in particular, can play an essential role in picking out the privileged reference frames needed to secure a consistent relational account of (kinematic) Cartesian motion.

Chapter 7 will examine a relatively overlooked, but significant, set of hypotheses that Descartes' utilized in developing his relational theory of motion. Although they will ultimately be demonstrated to be deficient in various ways, a thorough examination of Cartesian motion warrants a close study of their important content. In the final two chapters of the book, the insights gained from chapter 6 will be put to use, and/or tested, by examining two apparently contrasting approaches to the problem of constructing a Cartesian spacetime. In chapter 8, for instance, we shall explore the possibility of utilizing the specific predictions offered by one of Descartes' seven collision rules as the foundation for a consistent relationalist Cartesian dynamics. This adaptation of Descartes' theory owes its origin to the work of Huygens, and invokes the concept of a center-of-mass frame. From the perspective of this reference frame, all bodies collide and rebound without losing speed, an interaction that conserves quantity of motion. (Reference frames will form an important role in our discussions, as will be evident from Part I onward, since they provide the means by which a relationalist can attempt to coherently describe the motion of bodies without positing absolute concepts, such as absolute spatial position.) In short, if all the collisions depicted in Descartes' impact rules can be subsumed under his first rule, which is the only accurate one in the entire set of seven, than a Cartesian has a means of conserving the quantity of motion in all bodily collisions. This formulation of Cartesian science essentially accepts one of the key assumptions in Newton's argument, namely that a relationalist cannot unambiguously describe the spatial positions traversed by a moving body over an extended temporal period of time and an extended region of space. On a relationalist theory, "place" is relative to the configurations of material bodies; so, if the bodies constantly change their relative positions, as is the case in Descartes' matter-filled, plenum universe, then the places occupied by bodies cannot be ascertained over time (such as the motion of the planets around the sun). Yet, if one determines motion from the center-of-mass frame, which provides a temporary measure of the speed of two bodies during the brief instants and small spatial regions spanned by their collision, then a relationalist can (hopefully) successfully treat all bodily interactions without the need of a reference frame that describes bodily positions and speeds over a much larger period of time and region of space. The dynamics-based "rest" force procedure for upholding Descartes' natural laws and relational motion, as developed in chapter 6, will also endorse this project for locating temporarily fixed reference frames.

In chapter 9, however, we will take the opposite approach: that is, we will investigate the prospects of constructing fixed reference frames that can describe the motions and collisions of bodies over non-local regions of space and time. If Newton's assumption is incorrect, and some form of permanent reference frame can be located in Descartes' constantly changing plenum, then the relationalist can meaningfully discuss the speeds and quantities of motion of all material bodies without the need to adopt absolute notions of position or velocity. On this interpretation, furthermore, it will no longer be necessary (or beneficial) to use the specific predictions of Descartes' collision rules as the basis of Cartesian dynamics. In fact, this approach to Descartes' dynamics will basically forsake our earlier preoccupation with collisions in order to concentrate on the essential "interconnectedness" of all bodily motions in the Cartesian plenum. Specifically, since the plenum is entirely filled with material bodies, the movement of any one body entails the simultaneous displacement of a vast host of others, a phenomenon that does not easily lend itself to an analysis solely in terms of bodily collisions. In order to gain insights into the motions of bodies in a plenum, and possibly discover a means of constructing fixed reference frames in such an environment, we will examine the basic details of the modern theory of machine parts, also known as the "kinematics of mechanisms" theory. With the mechanics of gears as a blueprint, the viability of resolving the difficulties raised against Descartes will be explored. Yet, as will be evident throughout the book, we will demonstrate that, regardless of the apparent success (or failure) in devising a coherent relationalist dynamics, the various Cartesian theories examined in our study inevitably accept some form of Newton's supposition for a "geometrical background structure."

Finally, as mentioned above, a considerable portion of this book will center upon the historical context, and contemporary debates, surrounding the particular hypotheses and concepts that form Descartes' natural philosophy, especially his dynamic theories. In chapter 3, an overview of Descartes' scientific approach and his natural laws will lead us into a discussion of the ontological assumptions and commitments underlying the Cartesian concept of "force." Chapter 3 will mainly examine the concepts of space, time, and motion as specifically presented in Descartes' primary scientific treatises, with the contextual development of the theory of relational motion, however, receiving much of the attention (as well as the related concept of "inertial" motion, whether in a straight-line or circular form). In chapter 4, on the other hand, the idealized conditions that proceed the Cartesian collisions laws will constitute the basis of a intricate study of Descartes' views of solidity, rigidity, and the individuation of material bodies in a plenum (to name just a few). Chapter 4 will also devote a great deal of attention to the vexed problem of Cartesian material substance. The intricate role of the Cartesian conservation law for the "quantity of motion" (or size times speed), examined in chapter 5, will likewise form an important part of this investigation, since it will be shown that the traditional approaches to Cartesian physics have not adequately treated, or previously understood, the many facets of Descartes' complex notion of a conservation law. All in all, investigating the specific details of Descartes' scientific theses will prove an invaluable asset to this study; for, as will become clear, many insights into the possibility of creating a consistent relationalist dynamics can be acquired from this most important of sources.

As the reader will be reminded throughout the investigation, while there have been no shortage of attempts to encapsulate Newton's natural philosophy in the modern mathematical formalism, Descartes' theories have not been as fortunate—thus the present study marks a much needed corrective to the prevailing historical bias. That is, one of the goals of the book is to demonstrate the added philosophical insights that can be gained by employing the modern "spacetime" formalism. In particular, we can briefly summarize a few of the overall conclusions of our investigation: (1) while most commentators have often vaguely perceived that a strict form of relationalism is incompatible with the Cartesian natural laws, most commentators have nevertheless been unsuccessful in perceiving that there are "weaker" formulations of relationalism that may adequately ground Descartes' physics (thus prompting our investigation for the elusive privileged reference frames, in chapters 6 through 9, that can provide a sound basis for a relational spacetime); (2) most commentators have also overlooked the fundamental importance of Cartesian dynamics in resolving the kinematic obstacles of Descartes' handling of relational motion (which, as will be first discussed in chapter 5, is directly supported in a largely unknown, but crucially important, letter to H. More). Moreover, as will argued in Part III, Descartes' handling of the collision rules strongly suggests that Cartesian bodies have a capacity, via his conservation law, for determining the individual motions of colliding bodies, thus establishing the groundwork for the privileged reference frames as mandated by the weaker form of relationalism he implicitly accepts.

PART I

DESCARTES, NEWTON, AND THE ABSOLUTE/RELATIONAL SPACETIME DEBATE

CHAPTER 1

NEWTON'S *De Gravitatione* ARGUMENT AGAINST CARTESIAN DYNAMICS

This chapter, and chapter 2, explore the ontological and methodological evolution of Newtonian, Neo-Newtonian, and Leibnizian concepts of space and time. These concepts will form the foundational basis, and comparative starting point, for the eventual construction of a Cartesian spacetime in later chapters. However, in order to adequately ascertain the basic ingredients of a spacetime modeled on Descartes' physics, we will need to briefly examine the genesis of Newton's spatiotemporal views, since these beliefs constitute the backdrop of Newton's famous argument against Cartesian space and motion. A short sketch of the relational theory of space and motion postulated by Descartes, especially as viewed within the context of the Cartesian theory of dynamics,[1] will be presented in chapter 1, so that the source of Newton's criticisms (also presented in chapter 1) will become readily apparent. (An in-depth examination of Descartes' theory will form the subject of Part II.) As mentioned in the Introduction, it is important to trace the precise implications of Newton's argument, since each relationalist theory we will examine will constitute a different attempt to answer the same Newtonian challenge. In chapter 2, consequently, I shall provide an extensive analysis of the details and adequacy of the various Neo-Newtonian and relationalist theories of space and time developed in the wake of Newton's initial hypothesis. These latter theories warrant our attention, for they strive to correct the perceived deficiencies intrinsic to Newton's original formulation of absolute space and time. Finally, in the process of discussing the merits of both absolute and relational theories in chapter 2, I intend to put forth an argument claiming that, in a sense crucial to the analysis of the motions of bodies, there is something essentially correct with Newton's demand for absolute space. This argument, although presented somewhat tentatively here, will assume greater proportions and significance as the difficulties involved in shaping a consistent Cartesian dynamics become more apparent with each succeeding chapter. Ultimately, we will attempt to determine if a committed Cartesian can accept this "correct" aspect of Newton's argument while still remaining loyal to Descartes' conception of relational motion.

1.1. The Two Trends in Cartesian Natural Philosophy

Among Descartes' numerous conjectures on the nature of space and time, the relational theory advanced in the *Principles of Philosophy* (published in 1644) is by far his best known and most influential contribution to this ancient puzzle. Although a more in-depth discussion of these issues will arise in the following chapters (namely, Part II), Descartes essentially accepted the Aristotelian doctrine (favored among the scholastics) that "place" denotes the boundary between an object's surface and the surface of the material bodies contiguous with that object, while motion is "*the transfer [translatio] of one piece of matter or of one body, from the neighborhood [vicinia] of those bodies immediately contiguous to it and considered at rest, into the neighborhood of others.* (Pr II 25)"[2] Of course, these assertions make sense only when they are conjoined with two further Cartesian doctrines: (1) that a body's matter is "identical" with its spatial extension (although they are conceptually distinct, see Pr II 10), and (2) the whole of space is filled with matter (Pr II 16). Relationalism enters this picture in the following manner: provided a universal plenum (i.e., a universe completely packed with matter), and his definition of "place" as the relative boundary between contained and containing bodies, he reasons that the stipulation concerning whether it is the contained or containing body that is really "at rest" is purely arbitrary, since "we cannot conceive of the body AB being transported from the vicinity of the body CD without also understanding that the body CD is transported from the vicinity of the body AB (Pr II 29)"; and thus, "all the real and positive properties which are in moving bodies, and by virtue of which we say they move, are also found in those [bodies] contiguous to them, even though we consider the second group to be at rest. (Pr II 30)".

Basically, in the relational theory we are considering, "motion" and "rest" are only meaningful or significant when presented as a velocity or acceleration difference (or lack there of) among bodies. But, a velocity difference is ambiguous with regard to the assignment of *individual* component velocities. If, for example, the velocity difference among two sailing ships totals 25 knots, it may be the case that one ship is completely stationary while the other moves the stated amount, or that both ships are moving at some specific velocity whose combined total equals 25 (say, 15 and 10, respectively). In such a scenario, an "absolute" or "actual" determination of each individual object's state of rest or motion is just not possible. This theory, which has attracted many adherents both before and since Descartes (Aristotle, Ockham, Leibniz, Mach, etc.), displays a number of admirable features. Principally, since the only observable motions in our universe seem to be the relative velocity and acceleration differences among material bodies, relationalism nicely aligns its conceptual apparatus and ontology with the content of experience. As a result, relationalists need not concern themselves with such notions as "absolute rest" or "absolute velocity" that may prove empirically unverifiable.

Unfortunately, in both the *Principles of Philosophy* (and his earlier *The World*), Descartes advocates a series of laws on the nature of motion which not only appear to contradict this relational view, but which provided Newton with the model for what was eventually to become the focal point of his own laws of motion: "When a body is moving, even if its motion most often takes place along a curved line . . . , nevertheless each of its individual parts tends always to continue its motion along a straight line."[3] Finished in 1633 (eleven years before his *Principles of Philosophy*), *The World* constitutes Descartes' greatest contribution to the development of dynamics (the branch of mechanics that deals with the motions of bodies under the action of forces); for, as revealed in the quotation, Descartes offers one of the earliest known hypotheses identifying the phenomenon of centrifugal force as occasioned by the "tendency" of bodies to move (inertially) in straight lines.[4] Alongside his thoroughgoing Aristotelian relational account of motion, consequently, Descartes also supported the existence of uniform inertial motion (i.e., a velocity that is constant and does not change, or accelerate) as an integral component of his world view. The second law of motion in the *Principles,* for instance, is identical to the quotation provided above: "All motion is, of itself, along straight lines; and consequently, bodies which are moving in a circle always tend to move away from the center of the circle which they are describing (Pr II 39)".

As discussed, it is not possible in a relationalist theory to ascribe motion of fixed magnitude and direction to a single body. Yet, Descartes proposes an hypothesis on the nature of motion in *The World* which practically embraces the notion that bodies can exhibit determinate individual velocities: "As for me, I conceive of [motion] none except which is easier to conceive of than the lines of mathematicians: the motion by which bodies pass from one place to another and successively occupy all the spaces in between (Descartes 1979, 63)." This geometric and seemingly "Newtonian" analysis of motion could not be more removed from his subsequent relational allegations in the *Principles*(as described above). Remarkably, one also discovers in *The World* two famous philosophical doctrines that are seemingly synonymous with Cartesianism: the identification of spatial extension as the sole property of matter (35-36), and the rejection of a matter-less void space (20-21). Faced with these apparently contradictory assertions, it seems that one must admit the hidden influence of a geometric, non-Aristotelian factor in Descartes' intuitions about place and motion.[5]

This underlying tension in Descartes' theory of motion is a problem which inevitably arises in the course of any examination of Cartesian natural philosophy. Among the many factors that might have precipitated Descartes' inconsistency, the disclosure of Galileo's condemnation by the Inquisition for teaching anti-Aristotelian, anti-relationalist physics has been assigned a prominent role.[6] Regardless of it source, Descartes' mature views on the nature of space and the phenomenon of motion, as put forth in the *Principles of Philosophy* (and defended throughout the remainder of his life), incorporate a set of apparently non-relational *laws* of motion (originally presented in *The World*) alongside a relational *theory* of motion. Despite the repeated charges of "inconsistency" and "incoherence" that are inevitably leveled at this strange marriage of seemingly opposing doctrines, Descartes' theory presents a paradigm example of the difficulties inherent in any attempt to reconcile an Aristotelian theory of place and relational motion with a (dynamical) theory of material body interactions. In chapter 2, we shall return to the investigation of the Cartesian theory; but, in the remainder of this chapter, we need to examine the argument that Newton specifically designed to exploit the apparent dichotomy of purpose exhibited by Cartesian physics.

1.2. Newton's Argument Against Cartesian Relationalism

Isaac Newton explicates his conception of space and time primarily in two documents: the relatively unknown *De gravitatione et aequipondio fluidorum,*[7] and the monumental *Philosophiae naturalis principia mathematica.*[8] Of the two, *De gravitatione* is by far the most revealing; since, among other things, the belated introduction of this early paper into the philosophical literature has finally implicated Descartes, and not Leibniz, as the intended target of his arguments against the relational theory of space and time.[9] In *De gravitatione,* Newton criticizes the divergent trends in Descartes' thought by demonstrating their basic incompatibility in several important respects. To demonstrate this point, we will exclusively explore an argument which, besides criticizing Descartes, likewise discloses the conceptual foundation of Newton's own spatiotemporal views. After disclosing the details of Descartes' theory, Newton states:

> I say that thence it follows that a moving body has no determinate velocity and no definite line in which it moves. And, what is worse, that the velocity of a body moving without resistance cannot be said to be uniform, nor the line said to be straight in which its motion is accomplished. . . . But that this may be clear, it is first of all to be shown that when a certain motion is finished it is impossible, according to Descartes, to assign a place in which the body was at the beginning of the motion; And the reason is that according to Descartes the place cannot be defined or assigned except by the position of the surrounding bodies, and after the completion of a certain motion the position of the surrounding bodies no longer stays the same as it was before. . . . Truly there are no bodies in the world whose relative positions remain unchanged with the passage of time, and certainly none which do not move in the Cartesian sense: that is, which are neither transported from the vicinity of contiguous bodies nor are parts of other bodies so transferred. . . .
>
> Now as it is impossible to pick out the place in which a motion began, for this place no longer exists after the motion is completed, so the space passed over, having no beginning, can have no length; and hence, since velocity depends upon the distance passed over in a given time, it follows that the moving body can have no velocity, just as I wished to prove at first. Moreover, what was said of the beginning of the space passed over should be applied to all indeterminate points too;

and thus as the space has no beginning nor indeterminate parts it follows that there was no space passed over and thus no determinate motion, which was my second point. It follows indubitably that Cartesian motion is not motion, for it has no velocity, no definition, and there is no space or distance traversed by it. So it is necessary that the definition of places, and hence of local motion, be referred to some motionless thing such as extension alone or space in so far as it is seen to be truly distinct from bodies. (Newton 1962a, 129-131)

Newton puts forth a number of important claims in this crucial passage. In order to better reveal its basic assumptions and form, it would be best to analyze his argument by detailing each important step. In what follows, premises (1) through (5) are all assumptions:

(**1**) Descartes' law of inertial motion: All bodies tend to remain at rest or move in rectilinear paths at uniform velocity. This is a conjunction of Descartes' first and second laws (Pr II 37-39).

(**2**) Descartes' relational theory of place and motion: This premise entails that all places and motions are determined relative to other contiguous material bodies.

(**3**) Descartes' plenum theory of matter: All of space is filled with material bodies. This premise is not essential.

(**4**) Observation: All the material inhabitants of the universe constantly alter their relative positions (relative to one another).

(**5**) Both straight line motion and velocity (which is described as *distance* divided by time) require a temporally fixed path of determinate length: That is, in order to determine a body's velocity and line of motion, the places successively occupied by the moving body, which together comprise a definite length, must remain unaltered over time.

(**6**) From (2) through (5): Straight line motion and velocity cannot be determined in the Cartesian universe. More explicitly, due to the continuous motion and scattering of the contiguous bodies responsible for defining relative place, the trajectory or path of a moving object can exhibit no well defined length and, thus, no well defined velocity (or "speed", since Cartesian motion is a scalar quantity, and thus not coupled to a definite direction—see chapter 3).

(**7**) Contradiction from (1) and (6).

(**8**) Conclusion: Premise (2) must be false.

Of course, following the logic of the argument, it is perfectly consistent to claim that the contradiction results from premise (5): in other words, that (2) is true and (5) is false. (Alternatively, one could also claim that the contradiction results from premise (1), but this would undermine Newton's similar first law of motion, so this option is rejected outright by the Newtonian absolutist.) Notwithstanding the source of the contradiction, Newton resolutely concludes that the only viable means of coherently defining the velocity and trajectory of a moving body is through the adoption of "absolute" space and time (which will be discussed below). In short, he believes (2) must be replaced with:

(**2a**) Absolute space and time: All spatial positions or places are temporally fixed and determinate. Thus, the spatial trajectories or paths of bodies, which are comprised of individual places, do not alter or change over time.

Although Newton also intended to refute the Cartesian hypothesis of place and motion with his famous "bucket experiment,"[10] the argument quoted above is particularly insightful for spelling out in much greater detail a famous component of Newton's theory of space and time; namely, the idea that one can meaningfully discuss a fixed spatial position or place that endures over a temporal period. In order to *define* the length of a spatial path, argues Newton, one must be able to identify through time the initial spatial position from which the motion commenced, as well as the intermediate places through which the motion persists. If this cannot be accomplished, then the length of the spatial path—and, consequently, the velocity and direction of motion—cannot be meaningfully expressed. Despite being specifically directed against Descartes' specific theory of place (i.e., that the bodies contiguous with the outer surface of an object determine its place), Newton's argument would appear to pose a threat to virtually any relational theory of motion. For even if a relationalist relaxes the definition of place to allow distant objects to specify the initial place of a moving body (e.g., Mach's "fixed stars"), there could exist no guarantee that those objects responsible for the designation (of the initial place) would remain in the same relative spatial configuration during the temporal interval required for the object's motion. For Newton, once the bodies that provide these position identifications move, the designation of the moving body's initial place will be irretrievably lost (e.g., no longer defined). Thus, the velocity and motion of a moving object cannot be calculated by any relational theory given Newton's requirement for an "absolutely" determinable trajectory. Finally, it should be noted that Newton's argument incorporates some very subtle ontological and epistemological issues, most of which are more fully disclosed in his specific attacks on Descartes' vortex theory of planetary motion. Since an analysis of the vortex theory is required to examine these additional features of Newton's argument, we will postpone this discussion until a later chapter.

Overall, Newton's argument against relational motion tacitly acknowledges the influence of what may be called (for lack of a better word) a "geometric insight" into the structure of the physical world. Since relational theories cannot adequately account for the observed velocities of bodies, Newton deems it necessary to install or posit geometric entities which can fulfill this role; namely, structures in an absolute space and time. What is remarkable about this form of reasoning is not merely that it assumes a very strong brand of realism, but that the realism it accepts is highly "geometric" in nature. In other words, Newton views the universe as inherently possessing some sort of "real" or "existing" structure analogous to the mathematical and geometrical structures required to explicate motion. These geometrical structures figure prominently in, for instance, his understanding of velocity, since its definition relies on the distance traversed in a given time period. Consequently, in order to meaningfully explicate the observed phenomenon of inertial motion, which is comprised of uniform straight-line velocity, the universe must be constructed so as to allow the identification of spatial positions (and hence, lengths) over temporal intervals. Thus equipped, the structure or background framework of space and time mirrors the mathematical and geometrical structures used to define inertial motion. This geometrical background framework is the conceptual cornerstone of Newton's theory of space and time, and will be of importance throughout our examination of the absolute/relational controversy. In fact, as will be seen, one may attempt to avert many of the problems intrinsic to Newton's formulation of absolute space and time by recalling and utilizing his geometrical background supposition. In the following chapter, accordingly, we will investigate the specific details of spacetime theories properly equipped to accommodate Newton's hypothesis of inertial motion.

ENDNOTES

[1] As used in this study, "dynamics" refers to the science that investigates the interactions and motions of bodies under the action of forces, whereas "kinematics" will pertain to the study of bodily motions per se. In addition, as applied to the historical theories of Descartes, Newton, or other Early Modern philosophers, such terms as "dynamic", "inertial", "velocity", etc., are not intended to signify their precise meaning as employed within modern physics and mechanics; rather, these terms are meant to be applied, often quite roughly, to stand for the analogous concepts or ideas employed by these philosophers. Consequently, to talk about, say, Cartesian "force" or "velocity" is not to imply that Descartes understood and utilized these modern physical concepts; but, instead, that the modern concepts of "force"

and "velocity" can serve as approximate descriptions of analogous concepts in Descartes' physics.

[2] Descartes, 1983. Translations from the *Principles* are based on Miller and Miller but are checked against the Adam and Tannery edition of the *Oeuvres de Descartes* (1976). I will identify passages according to the standard convention: thus, Article 15, Part II, of the *Principles* will be labeled "Pr II 25." Passages from the French translation of 1647 will be prefaced by "Fr". Other translations that are based on the Adam and Tannery will be marked, "AT", followed by volume and page number (and with a footnote/parenthetical reference citing the source of the translation).

[3] R. Descartes, *The World,* trans. by M. S. Mahoney (New York: Abaris Books, 1979), 71.

[4] J. B. Barbour, *Absolute or Relative Motion?,* Vol. 1, *The Discovery of Dynamics* (Cambridge: Cambridge University Press, 1989), 425-432. However, it should be noted that the exact details of Descartes' analysis of centrifugal force are somewhat flawed. This will be explained in more detail in a later chapter.

[5] In addition, other comments in this work reveal that his identification of matter and space was by no means clear or free of confusion. For example, he states: "Each of [matters] parts always occupies a part of . . . space and is so proportioned to its size that it could not fill a larger one . . . (Descartes 1979, 53)." Of course, Descartes may only be speaking loosely or metaphorically in this passage. Yet, the combination of this claim with his geometric analysis of moving objects "occupying successive places" presents a strong case for an inner conflict of opposing intuitions.

[6] In fact, a letter has survived in which Descartes informs Mersenne, a fellow philosopher, that (in light of Galileo's censorship) he intends to suppress the publication of *The World* so as not to offend the Church. See M. S. Mahoney, introduction to *The World,* by R. Descartes, 1979, xii-xiii.

[7] (circa 1666-1670), trans. and eds. A. R. Hall and M. B. Hall, in *Unpublished Scientific Papers of Isaac Newton* (Cambridge: Cambridge University Press, 1962a), 90-156.

[8] In two places: the Scholium on space and time of the first edition, 1687; and in the General Scholium to the second edition, 1713. *Mathematical Principals of Natural Philosophy,* trans. A. Motte and F. Cajori (Berkeley: University of California Press, 1962b), 6-12, and 543-547, respectively.

[9] See, H. Stein, "Newtonian Space-Time", *Texas Quarterly,* 10 (1967), 174-200.

[10] By demonstrating that the manifest centrifugal force could not have been produced by the motion of the water relative to the sides of the bucket--its containing surface. See, R. Laymon, "Newton's Bucket Experiment", *Journal of the History of Philosophy,* 16 (1978), 399-413.

CHAPTER 2

THE STRUCTURE OF SPACETIME THEORIES

Having examined the general features of Cartesian space and motion, and Newton's famous criticism of this theory, we can now proceed to the analysis of the underlying theoretical, or structural, components of the theory of space and time presupposed in Newton's argument. This investigation will not only determine the extent of the deficiencies, if any, in Descartes' system, but it will also outline the necessary structural or theoretical remedies necessary to cure the Cartesian theory of its presumed deficiencies

2.1. Newtonian Space and Time

Rather than entertain suspect relational theories, Newton insists that a complete and comprehensive analysis of the phenomenon of motion must invoke the existence of absolute space and time. In *De gravitatione,* he provides a brief synopsis of this theory: "[space] is eternal, infinite, uncreated, uniform throughout, not in the least mobile, nor capable of inducing change of motion in bodies or change of thought in mind. . . (Newton 1962a, 145)." In the modern parlance, Newton's appeal to the uniformity of space would be construed in terms of symmetry requirements. That is, Newtonian spacetime is symmetric under spatial displacements (homogeneity) and reorientations (isotropy); or, quite simply, that all places and directions in absolute space are inherently similar in nature (more on this latter).[1] In addition to the conception of space as infinite, immovable, and incapable of effecting change in the motions of material bodies (all notions which would be questioned in the 20th Century), Newton also declares that the "moment of duration" is the same for all individual parts of space, or, in other words, that the totality of spaces experience an identical moment of temporal passage: "we do not ascribe various durations to the different parts of space, but say they all endure together. The moment of duration is the same at Rome and at London, on the Earth and on the stars, and throughout all the heavens (Newton 1962a, 137)." With respect to the material aspects of time, he states: "absolute, true, and mathematical time, of itself, and from its own nature, flows equably without relation to anything external (Newton 1962b, 6)." The uniformity of space is matched, consequently, by the regular and homogeneous passage of temporal instants. On this view, moreover, material objects cannot causally influence temporal succession; space and time are the arena, and not active participants, in the "drama" of material interaction. In essence, Newton envisions absolute space and absolute time as separate from the material contents of the world. As for the obvious question, “What are space and time according to Newton?” (substance, property, or something else?), we will return to this point in a later chapter.

Newton's picture of the nature of space and time is often presented in the modern scientific/geometric formalism as a Newtonian, or Full-Newtonian, spacetime (see, for example; Stein 1967, 174-176). That is, the apparatus of differential geometry is employed in an attempt to articulate the physical (and metaphysical) import of Newton's theory. One must not embrace these techniques unconditionally, though; since they can as easily detract from, as assist in, the analysis of Newton's concept of space and time. More specifically, the technical details of the modern spacetime models might obscure and render unintelligible many intended conceptual facets of Newton's natural philosophy. Nevertheless, if employed judiciously, such techniques can offer valuable insights into the underlying structures of spacetime theories.

Overall, if we envision spacetime as the four-dimensional totality of physical events, then Newtonian spacetime splits this structure into a three-dimensional "space" and a one-dimensional "time" each possessing a Euclidean metric (or distance function). This view of the physical world can be summarized as a "space plus time," $E^3 \times E^1$, or "enduring space". Thus, the specification of a body's position in the universe requires a pair of coordinate values: A fix of its spatial location and its moment in time. Since Newtonian spacetime uniquely separates all events into simultaneity classes (see Figure 1),[2] Newton's concept of space is represented by a series of three-dimensional planes or "slices," with each plane comprising an entire collection of simultaneous events. Hence, enduring space can be pictured as an infinite series of spatial planes (one plane for each simultaneous collection). Newtonian time is just the unique order and distance between these "planes of simultaneity" (via the operation of the temporal metric). Given this structure, all the events located on a slice A bear a fixed temporal distance between the events located on a slice B.

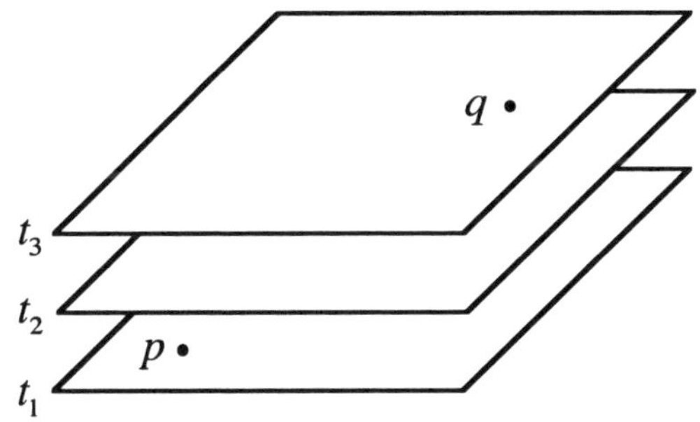

Figure 1. The planes of simultaneity represent the Euclidean three-dimensional space at each instant (in this diagram, p occurs at time t_1, and q at time t_3).

With respect to the geometry on each spatial slice, Newton insists that, "the positions, distances and local motions of bodies are to be referred to the parts of space (Newton 1962a, 137)," a claim that can be interpreted as an appeal to an intrinsic Euclidean spatial "metric" or distance function. In order to determine the actual or absolute (as opposed to relative) spatial lengths between objects and events, Newtonian spacetime thus requires that each point on the planes of simultaneity possess a "built-in" distance function. Essentially, a metric is a function which calculates the distance along a curve by summing over the infinitesimal lengths of the segments defined at each of the curve's points.[3] In Cartesian coordinates, these numbers are often expressed in the familiar Euclidean form $ds = \sqrt{dx^i dx^i}$, where ds signifies the (infinitesimal) length of the three-dimensional vector that connects the point x^i with the point $x^i + dx^i$. Furthermore, since Newton maintains that, "absolute space, in its own nature, without relation to anything external, remains always similar and immovable (Newton 1962b, 6)," the distance between events on each flat simultaneity plane is not conditioned by, or subject to, the altering and distorting influences of material bodies. Unlike the spacetime of General Relativity, where the distribution of matter actually determines the character of the metric geometry, the Newtonian metric is independent of the material objects and observers located in the spacetime.

As previously mentioned, however, one must proceed carefully when interpreting Newton's theory with the aid of modern geometric techniques; since such methods can present a somewhat misleading and distorted picture of Newton's actual intentions if not interpreted carefully. For instance, one should not construe Figure 1 as representing spacetime by a series of instantaneous spatial slices, since this is not only an incorrect interpretation of the figure, but of Newton's concept of substantival (absolute) space, as well. Figure 1 depicts $E^3 \times \{t\}$, for each specific moment of time t (i.e., t_1, t_2, t_3, etc.), and not $E^3 \times E^1$. The structure $E^3 \times \{t\}$ is useful for a graphic presentation of Newton's intuitions about the "moment of duration", but it can easily mislead one into thinking that $E^3 \times E^1$ is a union of individual spatial slices (which is, moreover, an misunderstanding of $E^3 \times \{t\}$). Consequently, Figure 1 would not seem to capture adequately Newton's overall concept of space as a persisting or enduring entity—it is a *single* thing which exists in or through time, not a collection of nearly identical things or spatial slices. His numerous claims on the status of absolute space, such as, "space is eternal in duration (Newton 1962a, 137)," would thus seem to corroborate the view that $E^3 \times E^1$ is the most natural modern-day geometric rendering of Newton's theory.

Notwithstanding these complications, however, the structure of Newtonian spacetime is generally quite useful in exhibiting other aspects of Newton's theory (as presented in the *De Gravitatione* argument). In particular, our four-dimensional approach can disclose the spacetime symmetries intrinsic to different methods of connecting or "stitching together" the instantaneous spatial slices. As explained, Newton obtains the velocity or directed speed of a body against the motionless backdrop of absolute space. Hence, his demand that spatial positions endure through time can be accommodated by isolating, or identifying within the larger structure $E^3 \times E^1$, the "fixed" points p of E^3 that form the set, $\{p\} \times E^1$. Each of the "lines" formed by these sets, $\{p\} \times E^1$, thus represents a point p of space that endures through time in Newton's sense, comprising a sort of spacetime "rigging" that cut the planes of simultaneity equidistantly in every slice.[4] Informally, the rigging can be pictured as a path or line which uniquely connects each spatial point with the same point on all the preceding and succeeding planes of simultaneity (see Figure 2). By way of this structural feature, one can meaningfully discuss in Newtonian spacetime whether or not a specific object occupies the same spatial location through time (since the rigging identifies the "same spatial location" on each spatial slice). Moreover, all objects (or events) in spacetime can be partitioned into one of two classes; those at rest with respect to absolute space (i.e., those that do not leave their spatial location), and those moving with respect to absolute space (i.e., those that do). As a result, all objects will possess "absolute velocities" and "absolute accelerations" relative to "motionless" absolute space. (The details of this process will be explained in the next section.) Moreover, it is important to note that our Newtonian spacetime is not equipped with an intrinsic geometrical structure which correlates or connects *in a unique manner* the one-dimensional time lines $\{p\} \times E^1$ with the flat Euclidean three-dimensional planes $E^3 \times \{t\}$. Put another way, many different sets of points $\{p\} \times E^1$ can be picked out in our Newtonian $E^3 \times E^1$ spacetime that stitch together, often rather differently, the $E^3 \times \{t\}$ spatial planes. A unique rigging thus cannot be derived from the structure and properties of the spatial slices and the temporal ordering: rather, a unique rigging must be incorporated into the our theory as an additional postulate, a fact that will assume importance below.

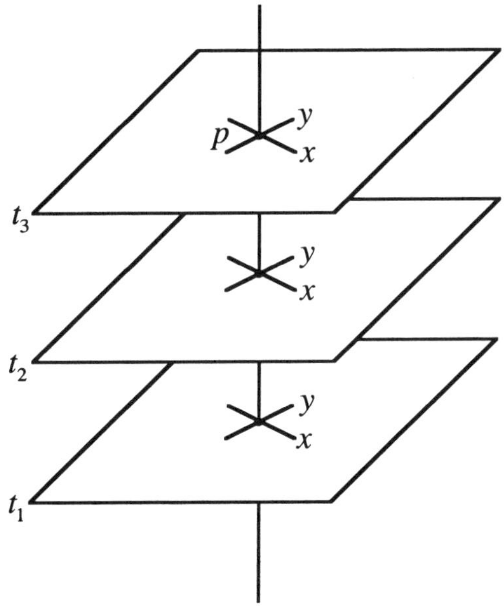

Figure 2. The spacetime rigging identifies the same spatial location p, which is the origin of a coordinate system (x, y), on each plane of simultaneity. The third spatial dimension of the coordinate frame has been repressed in this illustration.

Despite its great amount of structure, there remains a significant group of coordinate transformations that represent the inherent symmetries of our Newtonian spacetime. Briefly, there exists a class of coordinate structures or frames that can be characterized in terms of the given underlying structure of our Newtonian spacetime, and within this class of coordinates one can identify the coordinate transformations that "display" the inherent symmetries of that underlying spacetime structure. This interpretation of "symmetry" relies heavily on the concept of a "passive transformation" of coordinates, which is commonly understood as the functions that correlate the points (or places) of spacetime as viewed from one coordinate system to those very same points as viewed from a different coordinate system: thus, passive transformations merely re-label the same points of spacetime as viewed from different coordinates.[5] Accordingly, because these special coordinate transformations represent the inherent spacetime structure, the straight trajectories, absolute velocities, and absolute spatial locations of Newtonian spacetime do not change under these coordinate transformations. Once again, one must be careful not to misconstrue the intended purpose and implicit meaning underlying the use of our special coordinate transformations, however. The transformations merely *reveal* or *describe* the symmetries inherent in the spacetime structure: they do not *generate* those structures. Rather, it is the postulated connections and metrics (briefly described above) which determine the exact nature of the spacetime structure. Returning to the analysis of Newtonian spacetime, in rectangular Cartesian coordinates the spatial and temporal transformations from a point (x,t) in one reference frame to the same point labeled (x',t') in a different reference frame take the form:

$$x \rightarrow x' = R(x) + \text{constant} \qquad \text{(New)}$$
$$t \rightarrow t' = t + \text{constant}$$

The spatial transformations include an orthogonal matrix, $R(x)$, (or, simply, a series of correction numbers) that provide the coordinate frames with a constant rigid rotation. One can envision this process as the simultaneous or unison rotation of all the coordinate systems in spacetime. The constant term in the equation for spatial and temporal transformations describes the class of all translations. Briefly, this factor simply displaces or "moves," to an equal extent, all the coordinate systems. Provided a translation term, a spacetime can no longer retain a privileged origin or reference frame; since such schemes can be relocated at will both spatially and temporally. In fact, the spatial translations nicely demonstrate the symmetric properties of our Euclidean metric: no matter where we place the origin of our Cartesian coordinate system, the distance between two points remains invariant (as calculated by the function provided earlier, or, more simply, $\sqrt{(a_1 - b_1)^2 + (a_2 - b_2)^2}$ for points a and b in a two-dimensional space), although the coordinate values assigned individually to the points will vary from frame to frame. In conclusion, the coordinate transformations validate our original assertion that Newtonian spacetime is both homogeneous and isotropic (symmetric under spatial translations and reorientations—see Figure 3).

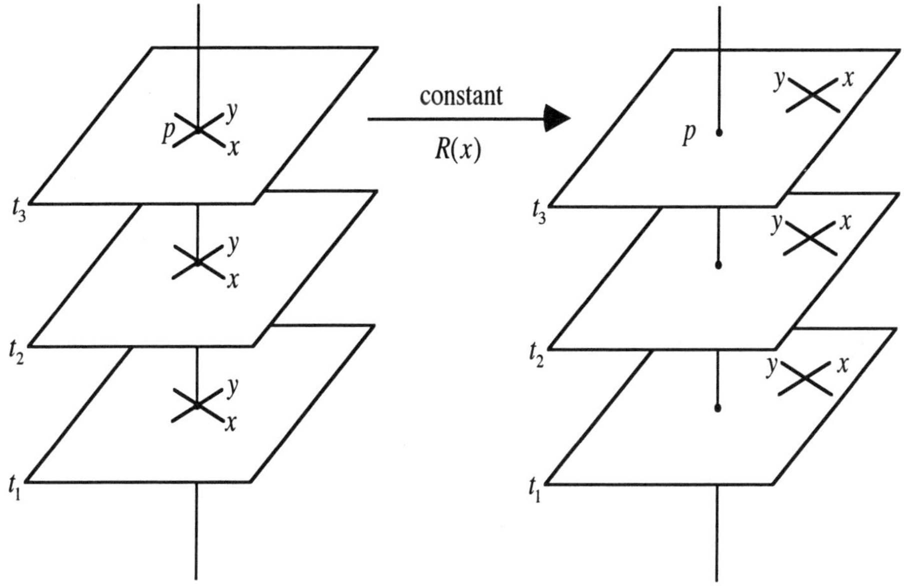

Figure 3. The spatial symmetries of Newtonian spacetime are comprised of all translations (constant), and all rigid rotations $R(x)$, of coordinate reference frames.

In his earlier *De Motu,*[6] Newton apparently endorses the view that the spatial translations form an important part of the group of spacetime symmetries: "Moreover the whole space of the planetary heavens either rests (as is commonly believed) or moves uniformly in a straight line, and hence the communal centre of gravity of the planets either rests or moves along with it. In both cases the relative motions of the planets are the same. . . (Newton 1962a, 301)." Since the entire "planetary heavens," including its center of gravity, can either remain at rest or move uniformly without disturbing the relative motions of the planets, Newtonian spacetime cannot possess a privileged origin or central point.

2.2. Galilean Relativity and Neo-Newtonian Spacetime

Yet, more importantly, the passage just quoted from *De Motu* reveals Newton's understanding of the deeper principle of Galilean relativity. In short, this principle holds that all inertial reference frames are physically equivalent or indistinguishable. The laws of physics and the behavior of material bodies, accordingly, will be identical in reference frames which are either at rest or moving with a constant velocity. On the basis of the Galilean principle, velocity only enters the laws of nature as a *relative* quantity; that is, as a "velocity difference" relative to a particular reference frame (e.g., "that ship is moving at 25 knots relative to our ship"). Newton provides a nearly identical definition of Galilean relativity in the passage quoted from *De Motu*: The relative motions of the planets are the same in a universe whose center of gravity (origin) either remains at rest or moves uniformly (in a straight line). Moreover, Newton incorporates this principle (also known as the restricted or classical principle of relativity) as a fundamental corollary in the *Principia*: *"The motions of bodies included in a given space* [i.e., reference frame] *are the same among themselves, whether that space is at rest, or moves uniformly forwards in a right line without any circular motion* (Newton 1962b, 20)."

In the third book of the *Principia,* nevertheless, Newton apparently retracts this analysis in a series of remarkable hypotheses and propositions: "Hypothesis 1: *That the centre of the system of the world is immovable.* Proposition 11, Theorem 11: *That the common centre of gravity of the earth, the sun, and all the planets, is immovable* (Newton 1962b, 419)." If we accept these statements at face value, and regard the *Principia* as Newton's last word on the foundations of the universe, then it would seem that we are forced to regard the center of gravity of the solar system (which he dubs "the center of the system of the world") as the *immovable* origin of space and time. Not only is the principle of Galilean relativity violated by these comments, because an individual reference frame has been assigned a determinate non-relative state of velocity (i.e., rest), but Newton would seem to have also installed within his spacetime a privileged origin. Hence, the symmetries of Newtonian spacetime would no longer retain the class of spatial displacements.

Yet, more charitably, since Newton furnishes an analysis of Galilean relativity in the *Principia* (see quote above, 1962b, 20) that is nearly identical to the passage cited from *De Motu,* one might construe Newton's "fixed origin" statements as pertaining to the position of the center of the universe *relative* to the orbiting planets. In other words, viewed relative to its rotating contents, the universe's common center of gravity is immovable. Thus, Newton presupposes an acceptance of the principle of Galilean relativity when he states that the center of gravity does not alter its position. By tacitly assuming this important distinction, Newton's further comments in the *Principia* can be successfully integrated into his overall project without seeming to imply that the universe incorporates the kind of privileged coordinate system that Aristotle favored.

Nonetheless, Newton's lack of clarity on this point has generated much confusion among commentators. J. B. Barbour, for example, believes that Newton actually overthrew his Galilean inclinations in favor of a sort of post-Copernican geocentricism (Barbour 1989, 643). In support of this view, one can point to the following incriminating remark from the "The System of the World", presumed to be an early draft of Book III of the *Principia:* "the common centre of gravity of all [the planets] will either be quiescent, or move uniformly forwards in a right line: in which case the whole system [of the planets] will likewise move uniformly forwards in right lines. But this is an hypothesis hardly to be admitted; and, therefore, setting it aside, that common centre will be quiescent: and from it the sun is never far removed (Newton 1962b, 574)." Newton offers this astonishing bit of reasoning when elaborating upon hypothesis 1 and proposition 11, theorem 11 quoted from the *Principia* above (1962b, 419). As presented, it is hard to determine the underlying motivation for Newton's dismissal of the potential Galilean motion of the solar system (possibly to appease the masses?): he merely states that, "this is an hypothesis hardly to be admitted," as if common sense necessarily rejects outright such absurd notions.

If we grant Newton a privileged origin in his spacetime, which essentially amounts to a fixed spatial location, then the constant term in the equation for spatial coordinate transformations must be dropped. As a result, the spacetime can no longer admit displacements of a coordinate system along any of the three spatial axes. We can represent this restriction of the allowable spacetime symmetries as follows:

$$x \rightarrow x' = R(x) \qquad \text{(Arist)}$$

It is hard to gauge the overall importance of this Aristotelian conception of space in Newton's natural philosophy. As explained, Galilean relativity is one of the foundational principles of Newton's theory of the physical world. Later, this issue will be examined at length when discussing the dynamical interactions of material bodies. However, as an aside, we can reasonably conclude that Newton's adoption of an Aristotelian fixed origin, if actually true, is both an unwarranted and unfortunate decision.

Generally, most commentators judge Newton's application of the principle of Galilean relativity to be inconsistent with his overall theory of absolute space and time. They argue, basically, that Newton's appeals to absolute rest and absolute velocity are purposeless or futile provided his further acknowledgment that such states of material bodies are empirically equivalent and indistinguishable. To borrow Michael Friedman's explanation, Newtonian spacetime draws a distinction between those bodies at rest and those moving with a constant velocity relative to absolute space, yet *the theory itself* does not furnish a means of distinguishing those individual states. (Friedman 1983, 276) It was this "over-determination" or unnecessary postulation of theoretical structure which led both Leibniz and Huygens to criticize Newton's theory.[7]

Nevertheless, the construction of spacetime models in the wake of modern mathematical advances has provided a means of retaining the essential features of Newton's picture without the need of an absolute velocity concept. In an ironic twist, the modern Newtonian can utilize advanced geometric techniques to undermine Newton's *own* argument for a fixed kinematical rigging. One can represent the elimination of the excessive structure of Newtonian spacetime by including a constant velocity factor, vt, in the equation for spatial coordinate transformations (the temporal equations remain the same). The Galilean spatial transformations now read:

$$x \rightarrow x' = R(x) + vt + \text{constant} \qquad \text{(Gal)}$$

In essence, the velocity term signifies the removal of the spacetime rigging that linked together the same spatial locations on the different planes of simultaneity. Neo-Newtonian spacetime, as it is called, thus allows all the coordinate frames to be simultaneously "sheered" or "tilted" by a fixed degree in a multitude of ways (depending on the value of the velocity term—see Figure 4). As a result, all the inertial (non-accelerating) particle trajectories or reference frames both at rest and moving with a constant velocity will be equivalent and indistinguishable, since there no longer exists a privileged class of trajectories—our unique sets of points $\{p\} \times \mathrm{E}^1$ (as above)—delineating the locations of stationary objects. The elimination of the "same place" rigging thus brings about a corresponding elimination of *a unique* identification of the same spatial location over time (across different spatial planes): given a point p on a given spatial slice $\mathrm{E}^3 \times \{t\}$ (for some moment of time t), any number of different inertial paths, or set of points $\{p\} \times \mathrm{E}^1$ (originating from diverse spatial points on the preceding planes), could pass through p. Basically, rather than entertain notions of absolute rest and absolute velocity via a rigging, Neo-Newtonian spacetime deems as meaningful only those questions relating to the *velocity difference* between inertial paths or reference frames.[8]

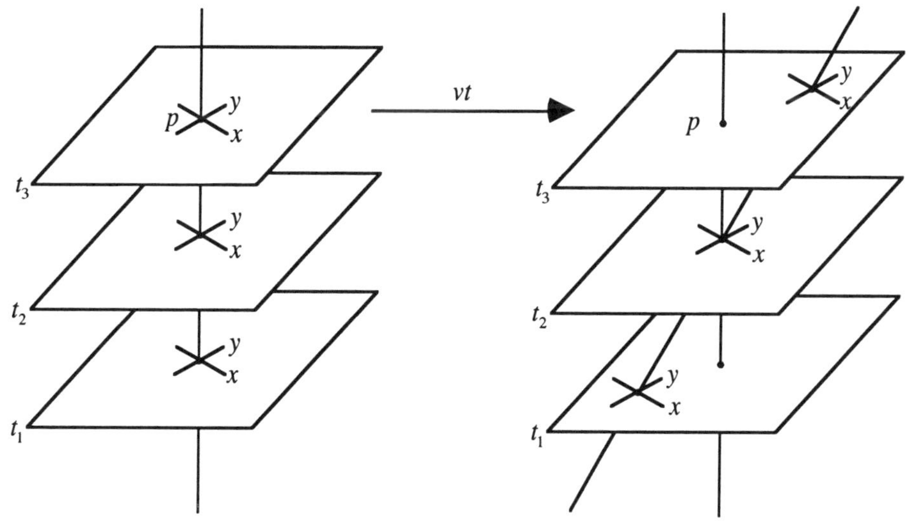

Figure 4. The Galilean transformations among inertial frames remove the spacetime rigging by allowing the coordinate systems to be "sheered" in numerous ways (depending on the value of vt). Note that the coordinate values are identical at t_2. Also, the affine lines that intersect t_2 are equally "straight:" The apparent perpendicularity of the line through p, as opposed to the other, is a mere artifact of our two-dimensional illustration.

In both Newtonian and Neo-Newtonian spacetime, the quantity "change in velocity" or acceleration can be determined without reference to the surrounding physical bodies. Yet, to measure this quantity in the latter theory, we must replace the single class of riggings in the Newtonian theory with the maximal families of riggings or lines generated by the Galilean transformations. These families of riggings are simply the possible inertial paths, or "affine" lines, through the spacetime (as described above). Given this structure, it is thus possible to determine which non-simultaneous events are inertial (relative to one another): all the events that can be linked by a possible inertial path of a particle, or by any of the lines parallel to this inertial path, form a class of inertially related events. In fact, these families of relatively parallel lines comprise our notion of an "inertial frame;" which is characterized by the absence of non-inertial forces (i.e., those forces associated with acceleration: centrifugal, gravitational, etc.) in any body at rest in that frame. Provided these inertial lines and frames, the average acceleration of a particle over a given interval of its trajectory can be calculated without recourse to the Full-Newtonian spacetime rigging. To accomplish this task, we first locate, on both the initial and last spatial slices (instants) of the interval, the three-dimensional displacement vectors that connect the affine line to the particle. Next, by dragging the displacement vector located on the final spatial slice back to the initial slice, a calculation of the relative displacement difference among these vectors, itself a new vector, provides a measure of "velocity" (or "change in displacement") over the given interval (see Figure 5). It is important to note, however, that the velocity measurements obtained from these processes are frame dependent: different inertial frames will arrive at different values of a single particle's velocity utilizing this system. To secure a physical quantity that *does* remain invariant across all inertial transformations, it is necessary to repeat the above process with respect to a particle's 3-velocity (i.e., the velocity vector on the three-dimensional spatial plane). If we drag the 3-velocity vector situated on the final time slice back to the initial slice of the interval, the difference between the velocity vector defined on that initial slice and final velocity vector (that we dragged back) supplies the three-dimensional "acceleration" vector, a quantity that retains the same value as viewed from all inertial reference frames. The frame-independent acceleration vector, which we can deem an "absolute" vector (for inertial transformations), has thus been derived from the frame-dependent measure of velocity.[9] While the value of the displacement and velocity vectors depends on the choice of inertial frame, the value of the acceleration vector does not; even though the former (displacement and velocity) are needed to formulate the latter (acceleration). Consequently, although Neo-Newtonian spacetime does not countenance the notion "absolute velocity," it does provide a measure of "absolute velocity change."

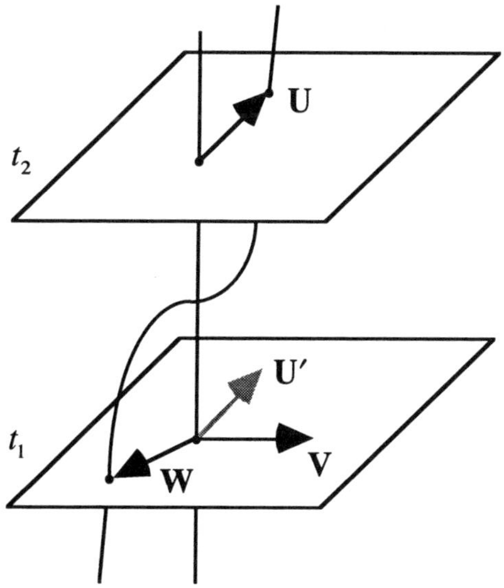

Figure 5. The affine lines allow determinations of the change in displacement by "dragging" the displacement vector $\mathbf{U}$ at t_2 back to t_1, labeled $\mathbf{U}'$, and comparing $\mathbf{U}'$ with the displacement vector $\mathbf{W}$ at t_1. The difference between these vectors ($\mathbf{U}' - \mathbf{W}$), a new vector $\mathbf{V}$, is the velocity over the given temporal interval. Repeating the process with respect to the velocity vectors provides the "absolute" acceleration vector.

2.3. A Frame-Independent Interpretation of Neo-Newtonian Spacetime

The conclusions of the previous section warrant further attention. In essence, a spacetime that incorporates the principle of Galilean relativity denies the existence of individual non-relative velocities. The measure of velocity is, therefore, determined relative to different frames of reference, which ensures that no single determination of that quantity can constitute the "true" measure of a given body's velocity. Due to the relative motion of the inertial frames, the estimation of an object's velocity will vary and differ from frame to frame. In the terminology of differential geometry, one would declare that velocity is not an invariant quantity of a moving body; where, as previously noted, "invariant" is defined as a quantity that retains the same numerical value under a certain group of coordinate transformations, in this case, our class of inertial coordinate systems related by our transformation rules. Nevertheless, the derivative of velocity, or acceleration, is an invariant property. Once the acceleration of a given body is determined in one frame, *it retains that same value in all other frames*. Accordingly, acceleration is an invariant quantity of our class of inertial coordinate systems; i.e., all frames will arrive at identical values when calculating a particle's acceleration.

Given this "absolute" frame-independent notion of acceleration, whose value is not dependent on the measurement of a unique inertial coordinate system, we can recover Newtonian dynamics without the need of an absolute velocity concept. In short, the basis or core of Newton's dynamical theory is his second law of motion, which usually takes the modern form "$F = ma$" (force equals mass times acceleration). This law must serve as the foundational element of all successful reconstructions of Newtonian dynamics, since it constitutes the primary differential equation of Newton's theory. As is readily observed, the second law only requires a determination of absolute acceleration, which does not automatically entail the need for an absolute measurement of velocity. Therefore, if we intend to eliminate frame-dependent quantities from the domain of natural laws, as most modern spacetime theories do, the mathematical facts would dictate that we include the 3-acceleration vector, as opposed to 3-velocity, in the construction of our Newtonian model. Frame-dependent quantities are often regarded with suspicion in the context of natural laws: if the laws truly apply to the entire universe, physical quantities that are valid only for a restricted class of observers, such as velocity, should not figure prominently in their construction. In fact, the law will generally hold only for observers in that restricted group of inertial reference frames (i.e., those with identical measures of velocity), thus defeating the purpose of its intended universality. What is desired for a universal law of nature is an invariant quantity, since it retains the same value for all observers (in this case, inertial observers). An appeal to frame-independent formulations of laws can therefore be viewed as a form of response to Newton's *De gravitatione* argument. Contrary to Newton's demands, a spacetime theory requires only the ability to determine absolute acceleration, and not absolute velocity, to explicate physical phenomena.

Nevertheless, reasons may exist to doubt the success of our frame-independent interpretation of Neo-Newtonian spacetime. Although only acceleration figures in the differential equations, it is still the case that the frame-dependent quantities of displacement and velocity are needed to calculate acceleration. This realization would seem to diminish the prospects for a coordinate-independent formulation of Newtonian dynamics: How can a theory claim to be frame-independent if its invariant quantities are definitionally dependent upon the existence of such frames? It was the understanding of these limitations in the construction of Newtonian and Neo-Newtonian spacetime models that lead, early in this century, to the development of a four-dimensional geometric approach. As will be demonstrated, these alternative renderings of Neo-Newtonian spacetime eschew the three-dimensional construction utilized thus far in this essay.

The key elements in this story, first developed by E. Cartan, can be presented briefly: rather than assign different metrics to both the three-dimensional spatial slices and one-dimensional time (i.e., the ordering of the slices), this version of the Neo-Newtonian theory installs a "degenerate" four-dimensional metric $\mathbf{g}*(\mathbf{V},\mathbf{V})$ on the entire four-dimensional spacetime (more on this below). Consequently, this metric operates on a four-dimensional velocity vector, a vectorial quantity that contains a temporal component in addition to the three spatial components employed in our earlier treatment. The 4-velocity vector of this new approach is a coordinate-independent geometric object: although its coordinate values will vary in different frames, the vector (a tangent to a parameterized curve) has an existence independent of coordinate systems and reference frames.[10] Hence, unlike our earlier construction, this version of Neo-Newtonian spacetime does not necessitate a frame-dependent quantity for the definition of its laws; in particular, Newton's second law.

In the differential geometry formulation of Neo-Newtonian spacetime, measurements of acceleration are procured by comparing the velocity vectors located at the boundary points of a given temporal interval, a procedure very similar to the frame-dependent technique outlined above. The "covariant derivative," as it is called, is likewise coordinate-independent geometric object, comprising, in fact, one of the most important structural features of our Neo-Newtonian spacetime (i.e., it is a structure "built-into" the very structure of the spacetime). The mathematical details of the workings of the covariant derivative are beyond the bounds of this investigation, but one can obtain an intuitive grasp of this device if we simply regard it as a process for determining the relative change in a (velocity) vector over a certain path: the covariant derivative obtains this measurement by "dragging back" the final vector to its initial point, computing the difference between these vectors, and representing this change by a new vector. This new vector corresponds to the acceleration of the particle along the path. (see, Sklar 1974, 204-205) Often, this process is provided the following mathematical representation: Given a 4-velocity vector field $\mathbf{V}$ defined along a curve parameterized in temporal units λ (or $dx^\alpha/d\lambda$, in a coordinate frame $x^\alpha(\lambda)$ mapped on to the curve), the covariant derivative of $\mathbf{V}$ along the curve is $\nabla_\mathbf{V}\mathbf{V}$ (which takes on the familiar form $d^2x^\alpha/d\lambda^2 + \Gamma^\alpha_{\mu\nu}(dx^\mu/d\lambda)(dx^\nu/d\lambda)$ in the coordinate frame). When the acceleration vector everywhere equals zero, $\nabla_\mathbf{V}\mathbf{V} = 0$, the particle follows an inertial path over the given time period; these paths are called "geodesics". Furthermore, the value of the covariant derivative is an invariant of our transformation group; thus all other reference frames will likewise view this trajectory (for infinitesimal distances and times) as the particle's unique inertial path (of zero acceleration).[11] This feature of the differential geometry rendering of Neo-Newtonian (and Newtonian) spacetime derives from the tensorial nature of the covariant derivative. Consequently, we recover the inertial or affine lines of our earlier Neo-Newtonian theory by ascertaining the four-dimensional curves that satisfy $\nabla_{\mathbf{V}}\mathbf{V} = 0$.

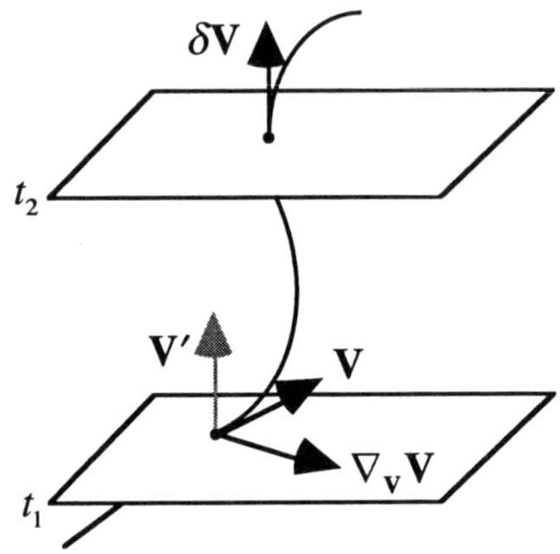

Figure 6. The covariant derivative determines the absolute change in the velocity vector by parallel "dragging" the vector $\delta\mathbf{V}$ at t_2 back to t_1, designated $\mathbf{V}'$, and computing the difference between $\mathbf{V}'$ and the velocity vector $\mathbf{V}$ at t_1. This difference constitutes the invariant acceleration vector $\nabla_{\mathbf{V}}\mathbf{V}$, a 3-vector which lies completely on the plane of simultaneity (as opposed to the velocity 4-vectors).

Of course, Newton did not find it necessary to provide a temporal component to his notion of velocity: that is, Newton conceived velocity spatially at an instant (on the planes of simultaneity). In these terms, Newton's argument presumed that a unique 3-velocity had to be defined on the simultaneity slices. Given a particle's 3-velocity, which is determined relative to the spacetime rigging, Newton derives the 3-acceleration by merely noting the change in the 3-velocity. Nevertheless, our Neo-Newtonian spacetime can recapture Newton's demand for a 3-acceleration by placing a special restriction on the employment of the 4-metric. In essence, the "degenerate" metric $\mathbf{g}*(\mathbf{V},\mathbf{V})$ forces the acceleration vector $\nabla_{\mathbf{V}}\mathbf{V}$, which is the derivative of the 4-velocity, to lie completely on the spatial slices. By restricting the acceleration vectors to the simultaneity planes (formally, $dt(\nabla_{\mathbf{V}}\mathbf{V}) = 0$), this process converts the four-dimensional velocity vector into a three-dimensional quantity by discharging its temporal component, a result which nicely accommodates Newton's requirement for a purely spatial definition of acceleration (see Figure 6). Moreover, as above, the subsequent 3-acceleration vectors that everywhere satisfy $\nabla_{\mathbf{V}}\mathbf{V} = 0$ (over a temporal interval) will represent the inertial paths through spacetime. Accordingly, if we employ the spatially restricted covariant derivative as a means of obtaining an invariant quantity of acceleration, a process that does not require frame-dependent measurements of displacement and velocity, then Newton's *De gravitatione* argument can be fortuitously bypassed or thwarted. Under this interpretation, Newton's second law can take the form $F = m(\nabla_{\mathbf{V}}\mathbf{V})$, thus securing the establishment of a coordinate-independent Newtonian dynamics without absolute notions of displacement and velocity.

Finally, the reader should bear in mind that our informal construction of a Neo-Newtonian spacetime, which started from considerations of frame-dependent quantities and lead to the postulation of frame-independent quantities, is only one method of constructing a Neo-Newtonian spacetime. One could have simply opted to investigate such spacetimes purely from a coordinate independent (and thus frame independent) standpoint. To present a rigorous mathematical formulation of all the variants of Newtonian spacetime is not the goal of this investigation.

2.4. Conclusions: Newton's Argument Updated

In retrospect, Newton's requirement that the spatiotemporal structure of the world discern the "absolute velocity" of all physical bodies is clearly too strong. All inertial frames are indistinguishable, hence the fixed rigging of the Full-Newtonian theory is an unnecessary and overly stringent postulation. Insofar as Newton's argument against the Cartesian relational theory of motion rests upon this assumption, it is not very successful. A means of designating a fixed spatial position over time is not necessary due to the "empirical inadequacy" of the absolute velocity concept. In other words, one need not require that space and time distinguish absolute velocity, since the property (if it can be deemed as such) plays no useful role in explaining the behavior of physical bodies. Therefore, returning to the analysis of his argument in section 1.2, what we labeled assumption (2a) must be deemed an unsatisfactory attempt to resolve the problem of inertial motion. However, one must not construe this judgment as a condemnation of the entire Newtonian argument against Cartesian dynamics; rather, it is only a criticism of one aspect of his theory.

In essence, the capacity to record a change in velocity (or acceleration) is the only requirement that a spacetime theory must meet; and, as mentioned above (section 1.1), a relational theory can account for such changes in a perfectly coherent manner. Utilizing Descartes relational theory, motion is the transference of an object from the vicinity of one set of contiguous bodies to a new set (and, once again, the relation is reciprocal). The Cartesian plenum thus allows one to measure *relative* velocity differences between various bodies, and, as a direct result, relative changes in velocity. Accordingly, at least with respect to its *basic* capacity to measure or determine relative accelerations, Newton's argument fails to undermine Cartesian relationalism.

Yet, Newton does not criticize Descartes' theory merely on the grounds that it cannot tolerate an "absolute velocity" concept (i.e., velocity relative to absolute space); Rather, Newton singles out the intrinsic relationalism of the *Principles* as incapable of explicating the uniform straight line velocity, or inertial motion proclaimed in Descartes' *Principles*. As discussed, inertial motion plays a crucial role in Cartesian natural philosophy, especially the collision rules (as will be more evident in the following chapters). Unfortunately, a theory of space and time that defines motion relative to the surrounding bodies, such as Descartes', will generally not present a set of reference frames which agree on the uniformity and direction of any given bodily motion. That is, since the contiguous matter enveloping a moving object provide the reference frames that track its motion, and since these surrounding bodies will invariably change their relative positions during the course of the movement (some as they "pushed" or displaced by the moving object), the vast majority of the reference frames will disagree on the direction and magnitude of the object's velocity. Consequently, due to the loss of a coherent “inertial motion” concept, Descartes' theory lacks a means of construing many of the important features of the dynamical systems he helped to develop! Interpreted along these lines, Newton's argument against the Cartesian relationalist theory of space and time (and most other relational theories) is quite powerful and effective.

Descartes' brand of relationalism is often presented in the modern guise of "Leibnizian spacetime". The main difference between Leibnizian and Neo-Newtonian spacetimes lies in the absence of an affine stitching or covariant derivative. Without this geometrical item, it is no longer possible to discuss inertial motion in an invariant or absolute sense—in fact, inertial motion really does not exist in Leibnizian spacetime. The spatial coordinate transformations that display the intrinsic symmetries of this spacetime are as follows:

$$x \to x' = R(t)(x) + a(t) + \text{constant} \qquad \text{(Leib)}$$

The added time variable of the rotation factor, $R(t)(x)$, signifies the newly created time-dependence of the rotation among coordinate reference frames.

Now, pairs of coordinate systems separated by any temporal interval can undergo varying degrees of rotation relative to one other. Likewise, in order to account for the phenomenon of Galilean relativity in this relational spacetime, the time-dependent function $a(t)$ replaces the constant vt (i.e., $a(t)$ is a function and not a constant, as is v). Overall, these time-dependent functions obliterate any chance of defining "straight-line" or inertial motion (see Figure 7).

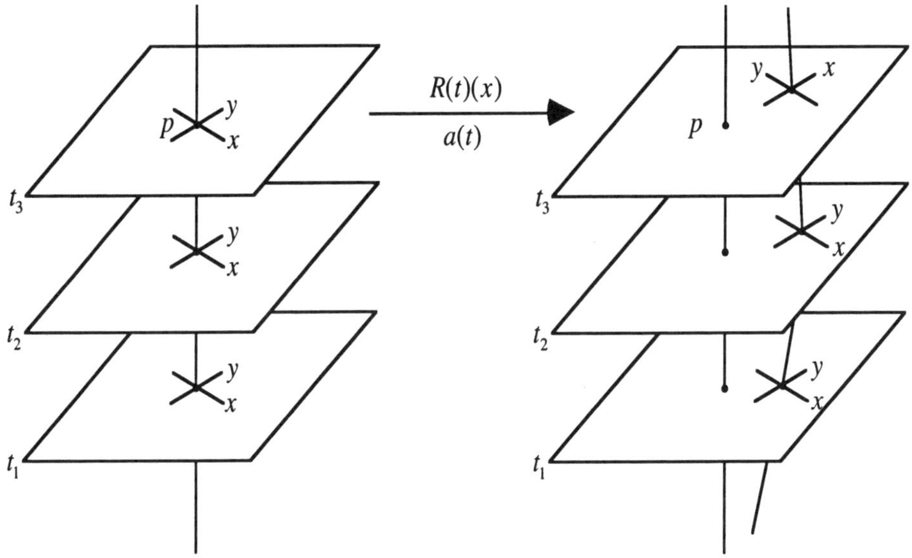

Figure 7. The time-dependent rotation matrix $R(t)(x)$ and Galilean function $a(t)$ form part of the coordinate transformations of Leibnizian spacetime (note the 90° rotation from t_2 to t_3—although this picture is somewhat exaggerated, since such functions must vary "smoothly" over time). Once again, the apparent "straightness" of the path through p, as opposed to the other trajectory, is only a feature of the diagram.

Returning to the postulation of an absolute theory of space and time, it is clear that something is essentially correct with what we earlier deemed Newton's intuition for a "background geometrical structure." As argued above, Newton firmly believed, with respect to absolute space and time, that the mathematical structures necessary for the elaboration of physical phenomenon are derived from, and grounded in, the ontological framework of the natural world. Thus, if the geometrical analysis of velocity requires unchanging spatial positions, then the physical world must likewise accommodate such structures. Yet, it can be convincingly argued that Newton failed to take into account all the relevant observational evidence regarding the phenomenon of velocity before invoking his physical apparatus. As his contemporaries Leibniz and Huygens were quick to point out (and Newton himself admitted in his postulates and corollaries in the *Principia*), all inertial reference frames are empirically and physically equivalent. Indeed, the velocity of objects with respect to absolute space plays no role in either the observed behavior of physical bodies or in the mathematics required to model such phenomena. Nevertheless, there are two factors which may have conspired to bring about Newton's apparent violation of his own empirical conclusions: (1) he recognized the need to equip space and time with the necessary structure to discern inertial motion; and (2), the lack of sophisticated techniques (described in this chapter) required to determine inertial motion (without recourse to a spacetime rigging) caused him to presume that a stronger "space plus time" structure was needed. In short, since Newton was intent on allocating the requisite geometrical structure to delineate inertial motion, and since "absolute spatial position" represented the best *available means* of reaching this goal, Newton would appear to have had no other choice but to violate the empirical import of Galilean relativity through the installation of this absolute structure. When Newton was faced with the dilemma of accepting either a theory too weak to render inertial motion coherent or one too strong to rationalize Galilean relativity, he chose the latter—concluding that too much is better than incoherent!

Essentially, given the laws of motion Newton inherited from Descartes, the capacity to "absolutely" distinguish uniform inertial motion from accelerated motion amounts to a prerequisite structure for an evaluation of many physical interactions. For example, the tendency of a rock to depart a rotating sling along a tangential path, which is exhibited by an outwardly directed force exerted on the sling, apparently validates the Cartesian/Newtonian contention that objects are naturally inclined towards inertial motion (and resist change from inertial states). In this respect, the inability of Leibnizian spacetime to provide trajectories of uniform direction and motion invariantly across all coordinate transformations renders the theory useless for Newton's purposes; since, returning to our example, some frames will view the rock as both accelerating and exerting a force, while others witness a force in the absence of motion (i.e., both the frame and the rock accelerate in unison, and thus appear mutually at rest relative to each other). In these latter frames, of course, the existence of an acceleration force in the absence of accelerated motion violates Newton's second law, $F = ma$. Moreover, and unlike Neo-Newtonian spacetime, there exists no affine lines or covariant derivatives—which we can generally dub "inertial connections"—to determine which reference frame is "actually" accelerating and which is "actually" inertial. In Leibnizian spacetime, whether or not a single particle follows a straight inertial path or a curved accelerated path depends purely on the choice of reference frame and will normally vary from frame to frame. Consequently, it will be impossible to link the non-inertial force of acceleration with a determinate class of trajectories as *an invariant feature of a class of privileged reference frames (or coordinate transformations).* In order for Newton's laws of motion to operate effectively in a spacetime setting, it is necessary that all observers associate the same forces with *the same class of motions or trajectories.* Leibnizian spacetime cannot fulfill this requirement since the same force will be linked with a multitude of straight and variably curved paths depending on the choice of reference frame.[12]

Of course, many relational theories attempt to overcome this problem by invoking a privileged reference frame or class of frames (e.g., the earth, the fixed stars) to delineate the inertial and non-inertial motions of bodies. For instance, Mach claimed that the non-inertial forces experienced by Newton's rotating bucket "*are* produced by [the bucket's] relative rotation with respect to the mass of the earth and the other celestial bodies."[13] Yet, this clever ploy merely reintroduces a truncated version of an inertial connection back into the spacetime arena. In these theories, the surrogate connection is no longer part of the structure of space and time, but is tied to a certain group of material bodies. Thus equipped, the spacetime as a whole, which includes events and material bodies, possesses the necessary geometrical structure to determine the inertially and non-inertially moving reference frames and particles. Nevertheless, this tactic, which is quite useful for the relationalist, only tends to confirm the adequacy of Newton's idea of a geometrical background structure: the existence of *some form of inertial connection,* no matter how restricted, is necessary to sufficiently capture the full content of our observational experience of the physical world (in this case, our experience of inertial motion). As presented in this context, we will define a "surrogate connection" as a means of transmitting information between spatial slices on the states of material bodies, especially information relating to accelerations and forces. It is not being claimed, consequently, that all relationalist theories employ a mathematical device identical to the covariant derivative (as defined above), rather, the contention is that all spacetime theories incorporating a system of physical laws require some method of connecting the information that exists on the spatial slices.

Therefore, in the setting of our modern spacetime models, Newton's argument amounts to a demand for the interslice structures needed to define acceleration. Physics wants to tell how the future states of bodies derive form their past states, but Descartes' theory of space and time, as well as Leibnizian spacetime, provide no coherent connection between them. This insight into the operation of physical theories constitutes Newton's implicit demand for a "geometrical background structure," and hence forms the heart of his anti-relationalist arguments. Interpreted along these lines, for the existence of a surrogate inertial connection in spacetime as a whole, Newton has presented an extremely powerful and convincing case. Nevertheless, in the remaining chapters of this book, mainly Part III, we will investigate the prospects of constructing a Cartesian spacetime that can meet Newton's challenge: that is, we will attempt to formulate a spacetime that can accommodate *both* the Newtonian demand for a "geometrical background structure" and the core content of Descartes' theory of space and motion.

ENDNOTES

[1] For a discussion of symmetry conditions, see, J. R. Lucas, *Space, Time, and Causality* (Oxford: Oxford University Press, 1984), 120.

[2] J. Earman, *World Enough and Space-Time* (Cambridge, Mass.,: MIT Press, 1989), 8. Much of the technical terminology and concepts will be drawn from Earman 1989.

[3] For a nice discussion of these details on a non-technical level, see, J. D. Norton, "Philosophy of Space and Time", in *Introduction to the Philosophy of Science,* eds. M. H. Salmon, et al. (Englewood Cliffs: Prentice Hall, 1992), 204.

[4] Here, my terminology is adopted from, M. Friedman, *Foundations of Space-Time Theories* (Princeton: Princeton University Press, 1983), 77.

[5] Additionally, the transformations can also be conceived as a structure preserving mapping on spacetime itself which takes the "old" points to "new" points as viewed from the same coordinate system (deemed "active transformations"). In this essay, however, I will exclusively represent the "passive" formulation.

[6] I. Newton, *De Motu,* in *Unpublished Scientific Papers of Isaac Newton,* trans. and eds. A. R. Hall and M. B. Hall (Cambridge: Cambridge University Press, 1962a).

[7] See, H. Stein, "Some Philosophical Prehistory of General Relativity", in *Minnesota Studies in the Philosophy of Science,* Vol. 8, eds. John Earman, et al. (Minneapolis: University of Minnesota Press, 1977), 3-49.

[8] L. Sklar, *Space, Time, and Spacetime* (Berkeley: University of California Press, 1974), 204-205.

[9] See, M. Wilson, "There's a Hole and a Bucket, Dear Leibniz", in *Midwest Studies in Philosophy Vol. XVIII, Philosophy of Science,* eds., P. A. French, T. E. Uehling, Jr., H. K. Wettstein (Notre Dame, Ind.: U. of Notre Dame Press, 1993), 211.

[10] See, for example; C. W. Misner, K. S. Thorne, J. A. Wheeler, *Gravitation* (San Francisco: W. H. Freeman, 1973), 48-50.

[11] Once the appropriate reference frame has been located where all the components of the metric tensor located at a point vanish (for very small regions around the point), one can determine the unique inertial path (or shortest line—geodesic) which advances temporally forward of the point (connecting the other points which also lie close along the geodesic). See, for example, D. F. Lawden, *An Introduction to Tensor Calculus, Relativity and Cosmology*. 3rd ed. (Chichester: John Wiley & Sons, 1982), 108-110.

[12] Nevertheless, this fact has not prevented modern relationalists from attempting to provide a relational basis for Newtonian mechanics via other means: e.g., J. B. Barbour and B. Bertotti, in "Gravity and Inertia in a Machian Framework," 1977, *Nuovo Cimento* 38B: 1-27. Barbour and Bertotti utilize action-at-a-distance principles to overcome the limitations imposed by relationalist spacetimes. Yet, it remains unclear whether such spacetime models can effectively explain the phenomena of non-inertial motion, especially rotation. See, Earman, ibid., 89-96.

[13] E. Mach, *The Science of Mechanics*. 9th edition (London: Open Court) 1942, although first published in 1883.

PART II

CARTESIAN PHYSICS

CHAPTER 3

THE CARTESIAN NATURAL LAWS

In chapter 1, it was necessary to briefly present Descartes' theory of space and relational motion in order to better grasp the motivation underlying Newton's argument against relationalism. If we intend to construct a Cartesian science immune to Newton's problem, however, an in-depth examination of the details of Descartes' natural philosophy is required. Only when all the components of the Cartesian theory have been revealed and their functions explained can the relationalist proceed to assemble a coherent version of Descartes' theory. Before we can effectively study, or even construct, a Cartesian spacetime, moreover, it is necessary to investigate the origin and specific content of his views on force and material interaction. These ideas represent a sort of framework or foundation on which a Cartesian spacetime must be built. Among these ideas, the Cartesian laws of nature figure prominently; for they form the basis of all applications of Descartes' relational theory of motion to the physical world. In this chapter, consequently, the content of the Cartesian natural laws will be analyzed in an attempt to uncover an effective means of resolving the dilemma imposed by Newton's argument (although the working-out of any promising candidates will have to await Part III).

3.1. The Laws of Motion

Foremost among the foundational principles of the Cartesian universe are the three laws of motion. As previously mentioned, Descartes' great contribution to the development of modern dynamics is his contention that moving bodies follow straight paths, an hypothesis that appears as the second law of nature in the *Principles*. Yet, one can also credit Descartes with the first classification of motion and rest as intrinsic or primitive *states* of material bodies without need of further explanation (although one must guard against the anachronistic interpretation that would credit Descartes with holding the modern concept of a "state" of motion). Thus, his first law of motion states "that each thing, as far as is in its power, always remains in the same state; and that consequently, when it is once moved, it always continues to move (Pr II 37)."

This realization, that a body remains in the same state unless acted upon by an external cause, is as important a conceptual breakthrough as Copernicus' situating the sun at the center of the Ptolemaic universe. For much of the Middle Ages, the Aristotle-influenced Scholastics endeavored to ascertain the causal principles responsible for the "violent" motion (or forced, unnatural motion) of corruptible, earthly bodies; that is, they focused their attention primarily on a category of momentary bodily movements on the surface of the earth that originate and conclude in a state of rest (in contrast to the perceived eternal and uniform rotation of the celestial spheres).[1] Given their lack of sophisticated mathematical and scientific devices for analyzing nature, it was probably inevitable that the Medieval philosophers would formulate the problem of violent motion as a quest for an agent or property temporarily possessed by moving bodies—thus, by their reckoning, the violent motion of all earthly bodies is occasioned by the intervention and retention of a sort of "pusher" property. The "impetus" theory suggested by John Philoponus in the sixth century A.D., and developed by Jean Buridan in the fourteenth century, is an example of this type of "qualitative" (as opposed to quantitative) theoretical reasoning. On the general outline of the impetus theories, these violent motions occur when a quality is directly transferred to a body from a moving or constrained source, say, from a stretched bow to the waiting arrow. This property generates the observed bodily motion until such time that it is completely exhausted or depleted, thus bringing about a cessation of the violent movement (and the arrow falls back to earth).[2] Implicit in the Scholastic view is the basic belief that a terrestrial body continuously resists change from a state of rest while situated upon the earth, since the depletion of the "pusher" property eventually effects a corresponding return of the body's original motionless, earthbound condition. This form of reasoning can be summarized in the following succinct question: "What causes and *keeps* a body in motion?"

Descartes, on the other hand, effectively bypassed this problem, for he instinctively accepted the existence of inertial motion (uniform or non-accelerating) as a natural bodily state alongside, and on equal footing with, the notion of bodily rest. He argues, "because experience seems to have proved it to us on many occasions, we are still inclined to believe that all movements cease by virtue of their own nature, or that bodies have a tendency towards rest. Yet this is assuredly in complete contradiction to the laws of nature; for rest is the opposite of movement, and nothing moves by virtue of its own nature towards its opposite or own destruction (Pr II 37)." Therefore, in contrast to the Scholastics, Descartes' conception that both uniform motion and rest are "primitive" facts (or basic, unreducible, etc., facts) of extended matter likely prompted him to develop his series of collision laws aimed at resolving the query: "What causes a change of motion (or rest)?" By posing the question in this manner, Descartes laid the foundation for the genuine breakthroughs in the study of motion that were to occur in the succeeding centuries (which is not to deny Galileo's immensely important role in this development). Where before the analysis had focused on explaining bodily movement, or velocity (as represented by the time derivative of the position function), the emphasis had now shifted to the description of change in motion, or acceleration (as captured by the time derivative of velocity). Yet, the importance of investigating acceleration, as opposed to velocity, was not immediately perceived in the seventeenth century. (As was noted in the previous chapter, even Newton overemphasized the significance of velocity by constructing an unnecessarily rigid spacetime in an attempt to determine the different states of inertial motion.)

While Descartes' first and second laws deal with the natural states of bodies mainly from the perspective of their individual non-interactive characteristics, the third law of motion is expressly designed to reveal the properties exhibited by collisions and interactions among several inertially moving bodies. In short, the third law addresses the behavior of bodies under the *normal* conditions in his matter-filled world: when they collide (which will be discussed at length below). "The third law: that a body, upon coming in contact with a stronger one, loses none of its motion; but that, upon coming in contact with a weaker one, it loses as much as it transfers to that weaker body (Pr II 40)." It is undoubtedly the case that Descartes has incorporated a form of conservation law within this postulate, but it is not yet clear which quantities, or possibly qualities, are being conserved. In the following sections of the *Principles,* Descartes makes explicit both the type and origin of his conservation law:

> We must however notice carefully at this time in what the force of each body to act against another or resist the action of that other consists: namely, in the single fact that each thing strives, as far as in its power, to remain in the same state, in accordance with the first law stated above. . . .This force must be measured not only by the size of the body in which it is, and by the [area of the] surface which separates this body from those around it; but also by the speed and nature of its movement, and by the different ways in which bodies come in contact with one another (Pr II 43).

As a consequence of his first law of motion, Descartes insists that the quantity conserved in collisions equals the sum of the individual products of size and speed of the impacting bodies. In some fashion, the size of a body corresponds to its volume and surface area, although we shall examine more closely the interrelationship of the concepts of volume and surface area in a later chapter. If we define B and C as the respective sizes of two bodies, and label their pre-collision inertial speeds v and w, and their post-collision uniform speeds v' and w', the equation reads:

$$Bv + Cw = Bv' + Cw' \qquad \text{(QM)}$$

This conserved property, which Descartes refers to indiscriminately as "motion" or "quantity of motion," is historically significant in that it marks one of the first quantitative attempts to come to grips with the problem of material interaction. In fact, Descartes envisions the conservation of quantity of motion as one of the fundamental governing principles of the entire cosmos. When God created the universe, he reasons, a certain finite amount of motion (quantity of motion) was transmitted to its material occupants; a quantity, moreover, that God continuously preserves at each succeeding moment.

> It is obvious that when God first created the world, He not only moved its parts in various ways, but also simultaneously caused some of the parts to push others and to transfer their motion to these others. So in now maintaining the world by the same action and with the same laws with which He created it, He conserves motion; not always contained in the same parts of matter, but transferred from some parts to others depending on the ways in which they come in contact (Pr II 62).

God's role in the Cartesian universe will be dealt with in the next section; hence, at this point, it will prove more profitable to closely examine the details of Descartes' conservation law. As developed in the *Principles,* it is important to note that Descartes defines quantity of motion as the product of size and uniform speed, and not size and velocity. Consequently, his conservation law only recognizes a body's degree of motion, which correlates to the scalar quantity "speed," rather than the vectorial notion "velocity" that pertains to a body's speed in a given direction (a distinction noted in chapter 1). This crucial distinction, between speed and velocity, surfaces in Descartes' seven rules of impact. Basically, Descartes found it necessary to augment his third law of motion with a series of postulates that spell out in precise detail the outcomes of bodily collisions (see Figure 8). A strict boundary is imposed upon their range, however, since the rules only describe the direct collisions between two bodies traveling along the same straight line (this problem will also be discussed at length in the next chapter). Nevertheless, Descartes' utilization of the concept of speed is clearly manifest throughout the rules. For example:

> Fourth, if the body C were entirely at rest, . . . and if C were slightly larger than B; the latter could never {have the force to} move C, no matter how great the speed at which B might approach C. Rather, B would be driven back by C in the opposite direction: because. . . a body which is at rest puts up more resistance to high speed than to low speed; and this increases in proportion to the differences in the speeds. Consequently, there would always be more force in C to resist than in B to drive, . . . (Fr Pr II 49).

Astonishingly, Descartes claims that a smaller body, regardless of its speed, can never move a larger stationary body. Leaving aside the obvious point that this rule is overwhelmingly disconfirmed by experience, the fourth collision rule demonstrates nicely the scalar nature of speed, as well as the primary importance of the quantity of motion, in Cartesian dynamics. In this rule, Descartes faces the problem of preserving the total quantity of motion in situations distinguished by the larger body's complete rest, and thus *zero value.* Without furnishing a rationale for his conclusion (at least in this section of the *Principles*—see below), Descartes conserves the joint quantity of motion by equipping the stationary object C with a resisting force sufficient to deflect the moving body B, a solution that does satisfy (QM) in cases where C is at rest.[3] That is, since B merely changes its direction of inertial motion, and not its size and speed, the total quantity of motion of the system is preserved: C equals zero throughout the interaction, so their combined quantity of motion is represented by the value of B. For Descartes, reversing the direction of B's motion does not alter the total quantity of motion, a conclusion that would seem to bear a certain amount of plausibility. This is in sharp contrast to the later hypothesis, usually associated with Newton and Leibniz, that regards a change in direction as a negation of the initial speed (from B to –B, a solution that, by contrast, is not nearly as intuitive). Thus, by failing to foresee the importance of conjoining direction and speed, which informs the concept of velocity, Descartes' law just falls short of that important breakthrough that would eventually lead to our modern understanding of the conservation of momentum.

In this context, the complex notion of "determination" should be briefly mentioned. Many passages in the Cartesian literature apparently refer to the direction of a body's motion as its determination: "there is a difference between motion considered in itself, and its determination in some direction; this difference makes it possible for the determination to be changed while the quantity of motion remains intact (Pr II 41)." As presented in this passage, the word "determination" seems to signify the direction of a given body's quantity of motion. Yet, Descartes takes Hobbes to task (through Mersenne) for making this very identification. In a letter dating from April 21, 1641, he states: "What he [Hobbes] goes on to say, namely that a 'motion has only one determination,' is just like my saying that an extended thing has only a single shape. Yet this does not prevent the shape being divided into several components, just as can be done with the determination of motion (AT III 356)."[4] Accordingly, just as a particular shape can be partitioned into diverse component figures, so a particular determination can be decomposed into various constituent directions. This notion is quite similar to the addition law of vector analysis, since a single determination can be conceptually broken down into a collection of several dissimilar determinations that originate from a common point. Given this distinction, one might plausibly define "determination" as the hypothetical composite direction of a body's quantity of motion.[5] In his *Optics,* published in 1637, Descartes seemingly endorses this interpretation during the course of deriving his law of refraction. He asks us to imagine the motion of a ball that is propelled downwards at a 45 degree angle, from left to right, through a thin linen sheet (see Figure 9). After the ball pierces the cloth, it continues to move to the right but now at an angle nearly horizontal with the sheet. Descartes reasons that this modification of direction (from the 45 degree angle to a smaller angle) is the net result of a reduction in the ball's downward determination through collision with the sheet, "while the one [determination] which was making the ball tend to the right must always remain the same as it was, because the sheet offers no opposition at all to the determination in this direction."[6]

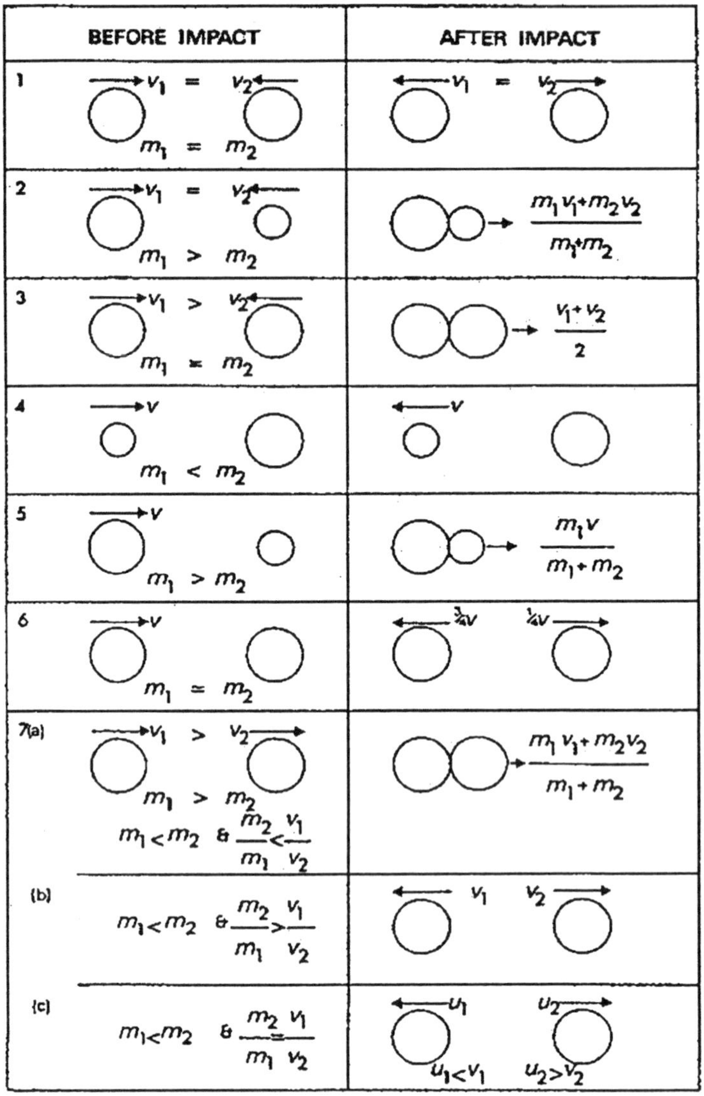

Figure 8. Descartes' seven collision rules (adapted from Aiton 1972, 36).

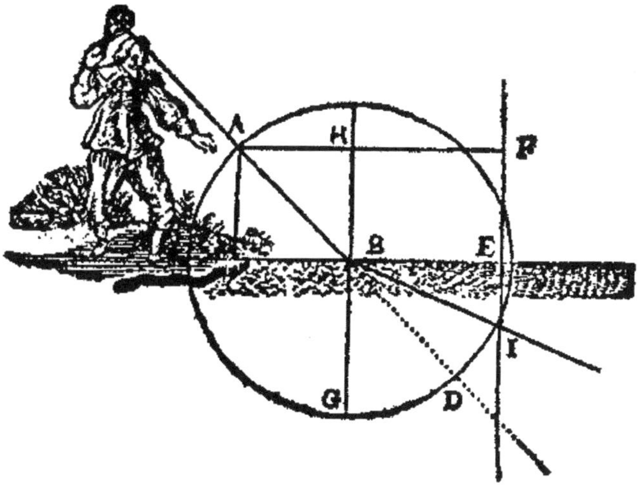

Figure 9. Descartes' diagram from the *Optics*. The angle of incidence of the ball before it penetrates the sheet ($ABC = 45°$) is greater than angle EBI after it passes ($EBI < 45°$) due to the decrease in downward determination.

Nevertheless, Descartes' determination hypothesis also incorporates a certain quantitative element. To demonstrate this feature of his dynamics, we need to return to our analysis of the fourth collision rule, which introduced the Cartesian force of resistance. Essentially, in order to account for a larger body's resistance to the motion of a smaller body, Descartes invokes the controversial thesis often described as the "principle of least modal action." In a famous letter to Clerselier (dated February 17th, 1645), Descartes explains:

> *When two bodies collide, and they contain incompatible modes,* [either different states of speed, or different determinations of motion] *then there must occur some change in these modes in order to make them compatible; but this change is always the least that may occur.* In other words, *if these modes can become compatible when a certain quantity of them is changed, then no larger quantity will change* (1991, 247).

This principle can be demonstrated with respect to our previous example. In order for both B and C to depart at the same speed and in the same direction after impact, it will be necessary for the smaller body B to transfer at least half of its quantity of motion to the larger stationary body C. Yet, Descartes reasons that it is easier for B in this situation to merely reverse it direction (i.e., a complete reflection) than to transfer its motion. He continues:

> When C is the larger [body], B cannot push it in front of itself unless it transfers to C more than half of its speed, together with more than half of its determination to travel from left to right [in this example, B moves from right to left], in so far as this determination is linked with its speed. Instead it rebounds without moving body C, and changes only its whole determination, which is a smaller change than the one that would come about from more than half of this determination together with more than half of its speed (1991, 247).

Consequently, reversing B's direction of motion, a change of one mode (determination), constitutes a lesser modal change than a transference of motion between two bodies, which alters two modes (speed and determination).[7] In this passage, it is important to note that if B were to transfer motion to C, it would change both half of B's speed *and* half of its determination, even though the direction of B's quantity of motion is *preserved.* As a result, a body's determination is not completely linked to the idea of direction: the former (determination) can vary despite fixing the latter (direction) "insofar as this determination is linked with its speed." Descartes has thus equipped his concept of determination with a quantitative dimension; a capacity apparently linked to a body's magnitude of speed. Once again, this determination hypothesis bears a close resemblance to modern mathematical notions; for a direction is now conjoined in some fashion with a scalar magnitude, as is the case with vectors. We will return to the discussion of determinations in the next section.

3.2. The Role of "Force" in Cartesian Natural Philosophy

Despite his efforts to encapsulate the laws of motion in a purely quantitative form, many tenets of Cartesian physics sit uneasily on the border between the qualitative speculation of the Scholastics and the mathematical inquiry of the later seventeenth century. In general, many of the concepts integral to the formulation of these natural laws are directly related to the central focus of this study—inertial motion and its accompanying forces. Thus, in order to fully comprehend the import of Cartesian space and motion, an examination of the conceptual details of Descartes' dynamical theory is essential.

As is well known, Descartes' devised his "mechanical" theory of the physical world in large part to reject and refute the widely held Scholastic explication of natural phenomena that employed an ontology of "substantial forms" and "primary matter." Briefly, the Medieval scholars, taking their initial cues from Aristotle, cultivated a doctrine that viewed material objects as comprised of both an inert property-less substratum (primary matter) and a determinate quality-bearing essence (substantial form). Hence, since matter is passive and thoroughly inactive on this account, it is only in the union with particular forms that matter is elevated to a causally efficacious status. A quantity of matter, for example, possesses weight, color, texture, etc., only when it is conjoined with a determinate form (of a billiard ball, chair, etc.).[8] In the *Sixth Replies* (to the *Meditations*), Descartes admits that he had earlier held such a view of gravity or "heaviness"; a view, moreover, that envisions the substantial forms in a teleological sense as a kind of goal-directed mental property: "what makes it especially clear that my idea of gravity was taken largely from the idea I had of the mind is the fact that I thought that gravity carried bodies towards the centre of the earth as if it had some knowledge of the centre within itself. For this surely could not happen without knowledge, and there can be any knowledge except in a mind. (AT VII 442)"[9] Descartes spoke out against this strange marriage of soul and matter in a remarkable passage from *The World* that essentially identifies the primary motivating factor in the construction of his mechanical system. He clearly and succinctly declares the Scholastic hypothesis to be both unintelligible and inadequate as a methodological approach to explaining natural phenomena:

> If you find it strange that I make no use of the qualities one calls heat, cold, moistness, and dryness. . . , as the philosophers [of the schools] do, I tell you that these qualities appear to me to be in need of explanation, and if I am not mistaken, not only these four qualities, but also all the others, and even all of the forms of inanimate bodies can be explained without having to assume anything else for this in their matter but motion, size, shape, and the arrangement of their parts (1979, 25-26).

Overall, Descartes' plan is quite simple: reduce the discussion of the class of non-observable metaphysically suspect properties, such as color, weight, taste, etc., to a discussion of the empirically sound observable properties of size, shape, and motion. In other words, Descartes intends to replace the teleological Scholastic hypothesis, which relied on a "mentally" influenced depiction of physical qualities, with a theory that requires only the primary properties of extension to describe the manifest order of the natural world. Consequently, *The World* constitutes the genesis of Descartes' search for a systematic explanation of all physical phenomena based solely on the operation and interaction of purely extended bodies.

Nevertheless, despite the strong mechanistic tendencies that Descartes displayed in analyzing the problem of motion in *The World* and the later *Principles,* the results obtained from his exhaustive study reveal a curious and intractable qualitative bent. In fact, returning to our earlier investigation of the specific character of the Cartesian laws, it is evident that Descartes has imbued his avowedly non-mental quantitative depiction of physical nature with many seemingly metaphysical and qualitative traits. A variety of examples are easily obtained: the tendency of bodies to follow straight lines, the resistance to motion evinced by a large resting body (to a smaller moving body), etc.. Accordingly, the formulation of his laws of motion concede a fundamental role to the action and existence of bodily forces or "tendencies." Many of Descartes' statements in both *The World* and the *Principles,* his chief scientific works, seem to verify this supposition; for example: "the virtue or power in a body to move itself can well pass wholly or partially to another body and thus no longer be in the first; but it cannot no longer exist in the world (1979, 15)." As an early remark concerning his conservation laws, Descartes tends to envision force much like a property or "power" possessed by individual material objects, although it is certainly the case that his language is far removed from the overt mind-influenced speculation of the scholastics (e.g., the impetus theory).

In addition, recalling our earlier discussion, Descartes' second law depicts inertial motion as an innate tendency of material bodies: "all movement is, of itself, along straight lines; and consequently, bodies which are moving in a circle always tends to move away from the center of the circle which they are describing (Pr II 39)." In order to better grasp the specific nature and implicit meaning in Descartes' use of force, it might be profitable at this juncture to closely examine his theory of centrifugal effects. At first glance, it seems as if the second law from the *Principles* corresponds to the standard scientific dissection of centrifugal force often encountered in school textbooks: on this modern view, the centrifugal effects experienced by a body moving in a circular path, such as a stone in a sling, are a normal consequence of the body's tendency to depart the circle along a straight *tangential* path.[10] Yet, as stated in his second law, Descartes contends (wrongly) that the body desires to follow a straight line away from the center of its circular trajectory. That is, the force exerted by our rotating stone, as experienced in the outward "pull" on the impeding sling, is a result of a striving towards straight line inertial motion directed radially outward from the center of the circle, rather than a striving towards straight line motion aimed along the circle's tangent.

Descartes does acknowledge, however, the significance of tangential motion in explicating centrifugal force,[11] but he relegates this phenomenon to the subordinate status of a composite effect. By his reckoning, the desire to follow a tangential path exhibited by a circling body, such as the flight of our stone upon release from the sling, can be constructed from two more basic or primary inclinations: first, the tendency of the object to continue along its circular path; and second, the desire of the object to travel along the radial line away from the center. It is the instantaneous composition of both the radial and circular tendencies that results in the observed inclination towards tangential motion. Thus, Descartes is willing to admit that "there can be strivings toward diverse movements in the same body at the same time (Pr III 57)," a judgement that seems to presuppose the acceptance of some type of "compositional" theory of tendencies analogous to his dissection of determinations. Yet, since he believes that "the sling, . . . , does not impede the striving [of the body along the circular path] (Pr III 57)," he eventually places sole responsibility for the production of the centrifugal force effects on the radially directed component of "striving."[12] In what follows, we need to define three points along a radial line extending outward from the midpoint of a circle: label E the center of the circle, A the point on the circumference, and D a point outside the circle (see Figure 10). He states, "If instead of considering all the forces of [a body's] motion, we pay attention, to only one part of it, the effect of which is hindered by the sling; . . . ; we shall say that the stone, when at point A, strives only [to move] toward D, or that it only attempts to recede from the center E along the straight line EAD (Pr III 57." Hence, the body's desire to move along the outward radial trajectory is the singular cause of the centrifugal pull on the sling, a force that is applied entirely along that outward radial path. (In the Appendix to this chapter, the "rotating sling" example, as we may call it, is furnished a more lengthy examination in order to assess the implications of Descartes' apparent invocation of a circular component of "striving".)

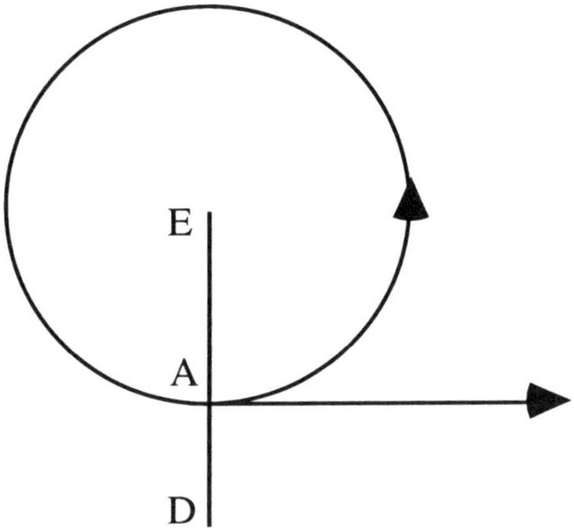

Figure 10. According to Descartes, a body in circular motion around the point E "strives" only to move along the radial line EAD when at A, rather than along the tangent to A.

It might be tempting to equate Descartes' use of the terms "tendency" and "striving" in his rotating sling example with his previous concept of a determination of motion. Despite their obvious similarities, especially with respect to their composite structure, the texts do not corroborate any such identification or correlation of concepts. As Garber correctly points out, a determination is confined to a body's actual motion (Garber 1992, 219-220). The concept of determination pertains to the direction of motion of a body undergoing a transference of place, and which includes all of its conceptually divided constituent parts. On the other hand, Descartes carefully couples the discussion of a body's tendency towards movement to the state of affairs presiding at a single instant. He states: "Of course, no movement is accomplished in an instant; yet it is obvious that every moving body, at any given moment in the course of its movement, is inclined to continue that movement in some direction in a straight line, . . . (Pr II 39)." In another passage in the *Principles,* Descartes identifies these strivings as a "first preparation for motion (Pr III 63)." Hence, while determinations necessitate a span of several instants, tendencies towards motion are manifest only at single instants. This is a crucial distinction, for it partitions Cartesian dynamics into two ontological camps: forces that exist at moments of time, and motions that can only subsist over the course of several temporal moments.

In reviewing Descartes' treatment of the concept of centrifugal force, the previously mentioned intrusion of qualitative explanatory elements is clearly evident. His depiction of centrifugal effects as a "tendency" or "striving" of a body possibly reveal a vestigial influence of his earlier Scholastic training. On the other hand, the importance of bodily tendencies in his physics could also be due to an awareness or intuition of the increasingly significant role that infinitesimal quantities had assumed in natural philosophy, a significance that would result in the development of the calculus in the late seventeenth century. Regardless of its origins (which are purely speculative), the unsought influx of qualitative elements into a quantitative scheme is not a problem endemic to just Descartes' natural philosophy. Many quantitative theories have been accused of integrating metaphysical concepts and ideas that allegedly offend certain branches of the scientific and philosophical communities: for example, Newton's adoption of a "force of attraction" (gravity) among material bodies was considered an illegitimate intrusion of an "occult" property by a host of Cartesian scientists (especially Huygens).[13] Thus, for our purposes, the important issue must concern Descartes' ambivalent attitude towards the existence of inertial forces, such as centrifugal force—an ambivalence all the more intriguing given his rejection of the Scholastic qualitative tradition. That is, even as his penchant for a geometrical world view increased, as manifest in the identification of extension as matter's primary quality or essence, Descartes continued to treat inertial motion and its accompanying force effects as if they were essential characteristics of bodies. Descartes' own remarks on ontological status of inertial force, moreover, bear witness to this explanatory inclination, as well as disclose a certain degree of ambiguity and indecision. In a letter to Mersenne dating from 1638 (six years before the *Principles*), he admits:

> I do not recognize any inertia or natural sluggishness in bodies. . . ; and I think that by simply walking, a man makes the entire mass of the earth move ever so slightly, since he is putting his weight now on one spot, now on another. All the same, I agree . . . that when the largest bodies (such as the largest ships) are pushed by a given force (such as a wind), they always move more slowly than others. (AT II 467)[14]

In this passage, Descartes seems to deny the existence of inertial force if conceived as a form of Scholastic occult quality that material bodies can possess or retain. Bodies are "inert" or "indifferent to motion," to use the standard terminology, so even the slightest weight should move the entire earth. On the other hand, he is willing to acknowledge the commonly observed fact that larger objects are much harder to set in motion than smaller objects. Consequently, although Descartes finds the existence of "forces of resistance" (or "natural sluggishness") problematic, as is the case with such similar properties as weight, he does not *entirely* relegate inertia to the phenomenological status of the so-called secondary properties of matter (such as color, taste, etc., which only exist in the mind). In some intricate or subtle manner, Descartes seems perfectly content to admit inertial force effects, or descriptions of these effects, into the domain of scientific discourse. Of course, what this concession amount to as regards ontology must remain a primary concern of Cartesian dynamics (as well as the problem of how inert matter can resist anything)?

All in all, I believe the underlying reason for the omission or exclusion, possibly intentional, of inertial force from the conceptual trash heap of occult properties can be found in Descartes' classification of motion as an intrinsic characteristic or "mode" of extension. As he had claimed in his earlier *The World,* the concluding sections of the *Principles* state: "I have now demonstrated [there] are nothing in the [material] objects other than . . . certain dispositions of size, figure and motion. . . (Pr IV 200)." Once again, these qualities—size, shape and motion—are accorded a special status due to their close relationship to extension. In effect, the modes of extension represent the means or manner by which something *is* extended. Thereupon, since inertial forces are a *form* or *consequence* of motion, Descartes apparently did not object to incorporating these phenomena within the discussion of the modes of material substance. This would also tend to explain his identification of the property "quantity of motion" as the force conserved in collisions, which he equates with the product of size (extension) and speed (motion). By regarding forces as a byproduct of the primary qualities of matter, Descartes rendered them palatable to his geometric dispositions.

Yet, even if Descartes described force as an intrinsic fact of material interactions, the exact nature of the relationship between force and matter remains rather unclear. In particular, is force a property actually contained or present within bodies? Or is it some sort of derivative phenomenal effect of the action of speed and size, and thus not present within extension? Recently, the role of force in Descartes' natural philosophy, and its place in his ontological domain, is a subject which has received much attention.[15] Although a thorough analysis of this problem would lead us too far astray of our present concerns—and would involve a lengthy discussion of the Cartesian conception of a "mode" and the Scholastic distinction between "causes of becoming" and "causes of being"—it is possible to briefly survey the principal sides in the debate.

On one reading of Descartes, Alan Gabbey's, forces exist in bodies in at least one important sense as "real" properties or modes whose presence occasions the Cartesian laws of nature. That is, the striking diversity and range of behavior commonly observed in bodily interactions, such as the behavior of colliding billiard balls, is due to the existence of inertial forces *within* the objects themselves (as a mode of body).[16] Overall, this analysis has the advantage of seeming to accord with many of Descartes' claims quoted in the previous sections. For example, in commenting upon his third law of nature, he states:

> A body which is joined to another has some power to resist being separated from it, while a body which is separate has some force to

remain separate. One which is at rest has force to remain at rest, and consequently to resist everything which can change it; while a moving body has some force to continue its motion, ... (Pr II 43)

Although the tenor of Descartes' explanation in this particular passage would seem to favor Gabbey's interpretation, that Cartesian force is an actual property possessed by bodies, Daniel Garber charges that such theories run counter to Descartes' demand that extension *alone* comprises the essence of matter. On Garber's reasoning, the existence of forces within bodies is just not compatible with Descartes' foundational claim: "Nor in fact does space, or internal place, differ from corporeal substance contained in it, except in the way we are accustomed to conceive of them. For in fact the extension in length, breadth, and depth which constitutes the space occupied by a body, is exactly the same as that which constitutes the body (Pr II 10)." Rather, Garber suggests that we view Cartesian force as a sort of shorthand description of the dynamical regularities maintained in the world by God: "The forces that enter into the discussion [of the Cartesian collision laws] can be regarded simply as ways of talking about how God acts, resulting in the lawlike behavior of bodies; force for proceeding and force of resisting are ways talking about how, . . . , God balances the persistence of the state of one body with that of another." (Garber 1992, 298) In the *Principles,* once again, it is possible to locate numerous passages where Descartes' so-called "divine sustenance" theory of the totality of universal motion seemingly supports this view: "So in now maintaining the world by the same action and with the same laws with which He [God] created it, He conserves motion; not always contained in the same parts of matter, but transferred from some parts to others depending on the ways in which they come in contact (Pr II 42)."

Generally, my sympathies lie with Garber's interpretation. I believe that his analysis of the texts best compliments my own earlier judgement that force constitutes a kind of "byproduct" or consequence of the motion of extended substance. Taking into account Descartes' persistent claim that extension constitutes the sole essence of matter, what I have described as force, or a byproduct of motion, should probably be deemed an apparent or phenomenal property of bodies, and not a property actually harbored within extension. Descartes' statements on the status of inertial force (quoted above) would seem to favor this interpretation, moreover. In that passage, as I have previously argued, Descartes apparently rejects the hypothesis that forces exist inside material bodies. The phenomenon of force, consequently, is simply the visible or experiential effect of the *ways bodies behave,* a pattern of action that is characterized mathematically as the product of their speed and size (and a judgment that leaves aside the question of its ultimate foundation—i.e., God). Although they are "inert," bodies act as if they house resistance forces. In retrospect, however, it must be acknowledged that Descartes' classification of material substance with extension, as exemplified in his demand that "[there] are nothing in the [material] objects other than . . . certain *dispositions* of size, figure and motion. . . . [my italics added] (Pr IV 199)," is so open-ended and equivocal as to easily accommodate both Gabbey's and Garber's hypotheses. In effect, can the product of two modes (or dispositions) of extension also qualify as a mode of extension?

In retrospect, although the concept of force is a fundamental concern of any venture to comprehend Descartes' theory of space and time, it must be acknowledged that his statements on the nature of these forces remain somewhat ambiguous. All that can be reliably concluded is that Descartes envisioned inertial force as a basic, possibly primitive, fact of the existence of material bodies; a broad, "safe" judgement that, by refusing to take sides, practically ignores the intricacies of the Gabbey/Garber debate. The spacetime absolutist may take comfort in this judgement, however. Without clear guidance from Descartes' texts, it would appear that the burden of harmonizing the Cartesian laws of nature (which includes the inertial forces) with a relational spacetime has been firmly and completely placed on the shoulders of the relationalist. In the next section, we shall finally begin to explore this relationalist dilemma.

3.3. The Cartesian Natural Laws and Relational Motion

As presented, the relationalist is faced with the difficult task of reconciling Descartes' avowed relationalism with the Cartesian dynamical theory of bodily interactions. This project's chief dilemma was briefly mentioned in the chapter 1: Is it possible to coherently expound velocity (or speed) and rectilinear motion in a world confined to the relative determinations of these quantities? In particular, the relationalist must accept Newton's challenge and provide a consistent description of nature from the sole perspective of material reference frames. The Cartesian relationalist is compelled to advance this project; since, by coupling the reference frames to the existing material bodies, Descartes elevated the notion of a "velocity difference (among bodies)" to a position of chief explanatory importance within his theory. Of course, one may claim that Descartes was unaware of this development, for he continued to portray his laws of motion in a non-relational quasi-“absolutist” manner despite his conversion to a form of Aristotelian relationalism. Accordingly, since an orthodox Cartesian must oppose this verdict, it is necessary to locate a theory or concept within Cartesian corpus that will finally integrate Descartes’ laws and material reference frames into a coherent and complete picture.

In essence, Newton contends that the concept of uniform rectilinear motion (inertial motion), or velocity, is incompatible with the specific character of Descartes' theory. If all the material bodies in the universe are in persistent motion, and motion is defined relative to these bodies, then a host of diverse trajectories and speeds will be legitimately attributed to the same moving (or non-moving?) objects. In surveying the damage to the Cartesian theory, one must admit that, without absolute spacetime positions or a fixed material reference frame, it is just not possible to salvage an intelligible relational description of inertial motion (which is derived from Descartes' first and second laws in the *Principles*). As argued in the previous chapter, a Cartesian relational universe can never admit absolute notions of "straight line" and "uniform speed." Hence, the relationalist project would seem best served by a construction or interpretation of the Cartesian laws that avoids or eschews reference to these meaningless notions; a strategy akin to the Neo-Newtonian dismissal of "absolute spatial position." Specifically, in order to retain the overall structure and purpose of Descartes' dynamical system, the Cartesian should locate and develop those laws and hypothesis within Descartes' texts that do not rely on the existence of uniform rectilinear motion. In implementing this scheme, of course, the Cartesian will need to modify many of Descartes' doctrines, and discard countless more. Yet, these actions can be justified on the grounds that the seed of a fully consistent and successful Cartesian dynamics lies concealed within the tangled mass of allegedly confusing hypotheses that comprise Descartes' theory. Although in a dormant state, they may claim, a proper "weeding out" of the extraneous and confused elements of Cartesian dynamics will finally allow this seed to take root and flower into a coherent physical theory.

On the whole, one might expect a close scrutiny of Descartes' third law of motion to reap the greatest relational benefits, since it seems to involve only the *relative* pre- and post-collision uniform speeds of two bodies along the same line. As stated in the first section, this conservation law is chiefly concerned with describing the interactions of several bodies: in contrast to his first and second laws, the third does not attempt to explain the *natural or intrinsic* behavior of *individual* bodies *outside of,* or *apart from,* a particular bodily collision. Unlike the first and second laws, the third law does not appear to imply an explanation that invariably introduces the concept of inertial motion as a basic fact of material existence. Provided this distinction, the uniform motions and straight lines employed in Descartes' third law can be regarded as merely accidental or temporary, and not essential, characteristics of *colliding* bodies. Thus, if the Cartesian could provide a purely relational account of the uniform speeds involved in bodily collisions (and the straight path), then Descartes' third law might form the basis of a fully coherent reconstruction of Cartesian dynamics.

Although the examination of the promising aspects of this approach will consume many of the remaining chapters, it can be quickly acknowledged that Descartes' own writings do not provide any straightforward or consistent support for this relationalist maneuver. Rather, Descartes usually depicts the bodily collisions that form this law in a decidedly non-relational fashion. In order to demonstrate this point, we will need to reexamine Descartes' collision rules. The fourth rule, as formerly mentioned, involves the interaction of a small moving body with a larger resting body. On his estimation, the larger object possesses a resisting force that the smaller body can never overcome. This force, while sufficient to deflect the moving body back along its initial path, guarantees that the larger body will remain in a state of rest during the collision with that smaller object. As noted, this rule is simply incorrect: small bodies can and do move larger resting bodies. Yet, the real problem for the relationalist concerns the conjunction of Descartes' fourth rule with his fifth rule:

> Fifth, if the body C were at rest and {even very slightly} smaller than B; then, no matter how slowly B might advance toward C, it would move C with it by transferring to C as much of its motion as would permit the two to travel subsequently at the same speed. (Fr Pr II 50)

In the collisions comprising this law, we are asked to consider the interaction of a small resting body with a larger moving body. On Descartes' reckoning, the large body will always move the smaller stationary object, transferring to it as much quantity of motion as is required to allow both bodies to travel at the same speed and along the same line while conserving the total quantity of motion. Once again, this rule does not hold in all real-world cases: these bodies will generally not move at the same speed or in the same direction after the collision.

Yet, for the relationalist, the contradiction in Descartes' analysis is painfully evident.[17] From the perspective of a relational theory, rules four and five constitute the *same* type of collision, i.e., they are an equivalent state of affairs, but Descartes draws completely different conclusions from each: in the fourth rule, one of the bodies reverses its direction after impact, while in the fifth rule, the collision brings about the joint motion of both bodies along the same course. Put another way, since a relationalist is confined to the notion of a "velocity (or speed) difference," it is not meaningful in such a theory to ascribe a component or individual state of motion (speed or rest) to each body—all that can be coherently discussed is the difference in motion among both bodies (see chapter I). In the present case, both rules present an identical scenario to the relationalist: a larger and a smaller body that manifest a variance in motion relative to one another. Therefore, because they represent the same physical events, a consistent relational theory must derive the same results from the situations outlined in collision rules four and five. The failure to foresee this simple relationalist fact must give the Cartesian cause to question Descartes' confessed devotion to the doctrines of Aristotelian relationalism.

Once more, in the *Principles,* we can observe the influence of a non-relational quasi-absolute conception of space in Descartes' handling of bodily interactions. This non-relational exposition, moreover, sheds light on a crucial distinction drawn in his first law of motion. According to Descartes, both rest and motion are intrinsic states of bodies without need of further explanation. Nonetheless, the evolution of his collision rules reveals that rest and motion are *distinct and opposing* states of bodies, a qualitative difference that cannot be captured by any relationalist means. That is, Descartes views a body's state of motion much like a contained property or quality: it is either actually "at rest" or actually "in motion." This distinction is apparent in the passage quoted earlier from his first law: "rest is the opposite of movement, and nothing moves by virtue of its own nature towards its opposite or its own destruction (Pr II 37)." Besides betraying the influence of the Aristotelian/Scholastic logic of contraries,[18] Descartes' pronouncement runs afoul of basic relationalist doctrine. On this spacetime theory, as just mentioned, there are no meaningful assignments of rest and motion to a single body. Only those description that involve a "velocity difference" among several bodies are sanctioned by the relationalist. Consequently, any attempt to construe the Cartesian collision rules from a relationalist point of view seems to contradict Descartes' own appraisal of the intrinsic bodily states of rest and motion.

In general, Descartes' tendency to portray motion in an absolutist, non-relational fashion renders most, if not all, reconstructions of his relational theory of motion untenable, or, at the very least, highly suspect. One may reasonably wonder, at this point, what the relationalist can effectively accomplish given such limited and ineffectual guidance from Descartes' texts? What strategy should the relationalist embrace in constructing a coherent Cartesian spacetime? In my opinion, the scheme outlined above represents a relationalist's best prospects: i.e., they need a version of Descartes' third law of motion stripped and shorn of its non-relational scaffolding, such that it does not appeal to the individual determinate motions of bodies outside of the context of particular collisions. This law, it would seem, is the key to a successful Cartesian dynamics. If successfully translated into the language of relationalism, the third law can surmount the obstacles put in place by Newton's argument. Part III will investigate the prospects of utilizing the third law of motion as the basis of a Cartesian dynamics.

3.4. Appendix: Circular Inertia

As discussed in section 3.2, the canonical presentation of Descartes' second law of motion (wherein "bodies that are moving in a circle always tends to move away from the center of the circle which they are describing," Pr II 39) would seem to invoke a tendency towards rectilinear motion within the context of larger, circular, non-inertial motion (i.e., non-inertial, or accelerating, according to the modern interpretation). Not only does this additional insight cast Descartes' mechanics in a different light, but it also greatly complicates any tendency to read Cartesian inertial motion as a straightforward prototype for Newton's system of natural laws. Fanning the flames of this general mystery is Descartes' influential explanation of the "rotating sling", in the *World* and the *Principles of Philosophy* (e.g., AT XI 85-86, Pr III 57-59), which seems to appeal to the rotating rock's "striving" towards circular motion—a concept that several commentators have taken as a latent form of natural— i.e., inertial—circular motion. In this section, we will explore the phenomenon of Cartesian circular motion in order to discern what role, if any, it plays in Descartes' physics. Contrary to those commentators who have pointed to the rotating sling examples as proof of Descartes' unfaithfulness to straight-line inertial motion (mainly, Westfall and Shea), it will be argued that a much more incriminating piece of evidence can be found in a largely-neglected letter from 1638 (to Ciermans, AT II 74). Moreover, after presenting the contrasting approaches to circular inertial motion in Galileo's and Beeckman's work, it will be argued that Descartes' utilization of circular inertia owes more to the philosophical ideas of the latter, as opposed to the former, which is a fact that has also been overlooked by commentators.

3.4.1. Circular Motion and Circular Tendencies.

The concept of circular motion has played a significant role throughout the history of science. Besides its integral importance to ancient astronomy, it figured prominently in the work of many early modern scientists on the inertial motions of both terrestrial and celestial bodies.[19] For our purposes, Galileo and I. Beeckman will be the subject of our investigation, since these two natural philosophers had a direct and powerful influence on the development of Descartes' physics.[20] The views of both of these philosophers will be presented before assessing the Cartesian theory, for it will later be argued that recent scholarship has largely overlooked the importance of Beeckman's contribution, possibly as a result of overemphasizing Galileo's influence. (Once again, as with Descartes, to employ the term "inertial" in this context is rather misleading, but we will retain its use in order to label those aspects of their respective hypotheses which resemble the modern conception.)

For Galileo, the circular motion of, say, a body orbiting the earth is a "neutral" state compounded of a center-seeking vertical motion (or tendency towards motion) and an impressed, usually horizontally-directed motion (or tendency towards motion). In brief, the Galilean analysis of circular motion incorporated hypotheses central to the Scholastic tradition: a quasi-Aristotelean "natural" motion towards the earth's center (which is governed by his "time squared" law for free-fall), and a non-natural (non-vertical) motion closely akin to the earlier "impetus" theories (see section 3.1). Although the details of his theory are somewhat vague, Galileo seems to hold that the impressed non-vertical motion not only balances out the vertical natural motion, but also accounts for the circular motion. He is much clearer on one point, however, once the vertical tendency of the body to descend has been held in check, any horizontal motion imparted to the body will result in the observed circular orbit of the body; and this circular motion will be "inertial" in the sense that it will persevere undiminished in the absence of the resistance of a medium. In the his early *Letters on Sunspots* (1613), he states:

> I have observed that physical bodies have an inclination toward some motion, as heavy bodies downward, which motion is exercised by them through an intrinsic property . . . . Finally, to some movements they are indifferent, as are heavy bodies to horizontal motion, to which they have neither inclination or . . . repugnance. And, therefore, all external impediments being removed, a heavy body on a spherical surface concentric with the surface of the earth will be indifferent to rest or to movement toward any part of the horizon. . . . Thus a ship, for instance, having once received some impetus through the tranquil sea, would move continually around our globe without ever stopping; and

placed at rest it would perpetually remain at rest, if in the first case all extrinsic impediments could be removed, and in the second case no external cause of motion were added.[21]

As noted above, the precise details of Galileo's theory are much debated by commentators, but his "qualitative" treatment of uniform circular motions (or tendencies towards motion) would find an analogue in Descartes' treatment of the problem (as will be discussed below).

I. Beeckman's approach to circular motion, on the other hand, marked a real advance on the road to the modern conception of inertia, since he reckoned that a uniformly moving body did not require a causal agent, or impetus, to explain its state of motion.

> Once moved objects never come to rest, unless they are impeded. Once an object is put in motion it never comes to rest, except by an external impediment. Furthermore, the weaker the impediment, by this will the object's motion be of greater duration; truly, if it is simultaneously projected high and moved circularly, it is evident to the senses that it does not come to rest before returning to earth; and if it were to come to rest, that it would not result from an equable impediment, but as a result of an inequable impediment because many parts of the air touch the moving object in succession.[22]

What is interesting about Beeckman's hypothesis, as can be discerned in this passage, is that inertial motion is not restricted to a linear path: any motion, even circular, qualifies as inertial motion. Only an external impediment can change/alter an internal "state" of motion, which entails that a body circling the earth in the absence of any resistance will never slow down or stop (or change its trajectory). By contrast, Galileo held that it is the imparted (tangentially-directed) impetus which, when accompanied by the center-seeking natural motion, guarantees the uniform circular motion of the body around the earth (also in the absence of resistance forces, of course).

Descartes' analysis of circular motion would seem to owe something to both Galileo and Beeckman, although the specific influences, if any, are not disclosed in his writings. In assessing this theory, we will need to recall Descartes' second law of motion, which states: "all movement is, of itself, along straight lines; and consequently, bodies which are moving in a circle always tends to move away from the center of the circle which they are describing (Pr II 39)." As first mentioned in section 3.2., it might seem that this law (from the *Principles*) incorporates the modern notion of centrifugal force: i.e., that the centrifugal effects experienced by a body moving in a circular path, such as a stone in a sling, are a normal consequence of the body's tendency to depart the circle along a straight *tangential* path (i.e., the stone, when released, will depart the sling and move along a path perpendicular to the sling's cord at the moment of release) Yet, as stated in his second law, Descartes reasons that the body desires to follow a straight line directly away from the center of its circular trajectory, and that the radially directed outward "pull" on the impeding sling is a result of this striving towards straight line inertial motion. (see Figure 10). Tangential motion, on Descartes' reckoning, thereby assume the status of a composite effect, since the tangential path exhibited by a circling body, as in our "rotating sling" example, can be constructed from two more basic or primary inclinations: first, the tendency of the object to continue along its circular path; and second, the desire of the object to travel along the radial line away from the center of rotation. Once again, it is the instantaneous composition of both the radial and circular "tendencies" (see section 3.2.) that results in the ensuing tangential flight of the stone upon release. According to the standard Newtonian explanation, the straight-line inertial tendency of the stone is combined, or balanced, by the person's "centripetal" pull or tug on the line, thereby bringing about the stone's circular motion (when in the sling). Descartes, on the other hand, reverses this explanation by using the combined radially-directed and circularly-directed tendencies towards motion *to be the cause* of the stone's observed tangential flight when released from the sling.

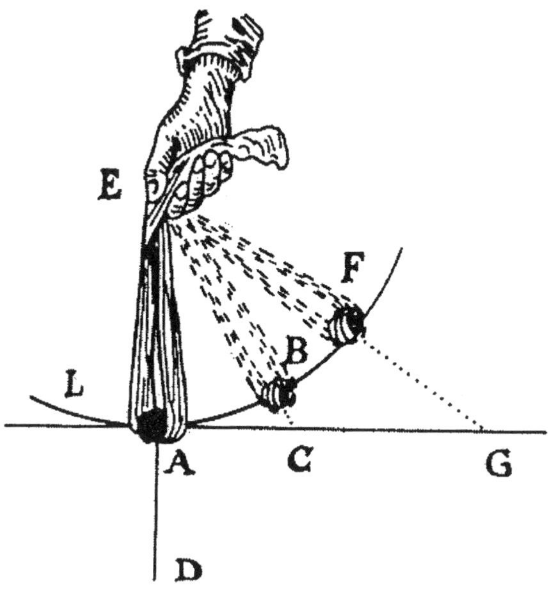

Figure 11. According to Descartes, a body in circular motion around the point E "strives" to move along the circular path to B and along the radial line EAD when at the point A.

For example, in Pr III 57, he insists that "the stone A, when rotated in the sling EA, definitely tends from A towards B (see Figure 11)," but "if instead of considering all the forces of [a body's] motion, we pay attention, to only one part of it, the effect of which is hindered by the sling; . . . ; we shall say that the stone, when at point A, strives only to move towards D, or that it only attempts to recede from the center E along the straight line EAD." By combining these two tendencies, he therefore concludes that if the stone "were to emerge from the sling when it reached the point A, it would go towards C and not towards B; and the sling, though it impedes this effect, does not impede the striving towards C." Thus, Descartes is willing to admit that "there can be strivings toward diverse movements in the same body at the same time," (Pr III 57). A similar dissection of circular motion is included in his earlier, *The World,* where he asks the reader to imagine the "inclination of the stone to move from A to C as if it was composed of two others, namely, one to move around the circle AB, and the other to move along the straight line [EGH, to match Figure 1] . . . ." (AT XI 85)

3.4.2. Critical Assessment of Cartesian Circular Inertia.

In critically analyzing Descartes' explanation of the rotating sling, it is important to bear in mind that he does not ascribe a circular inertial *motion* to the rock. Instead, the "inertial" implications of Descartes' analysis would appear to involve his assignment of a circular component of the body's "striving" towards motion (which, when combined with the radially-directed outward striving, gives rise to the tangentially directed, composite striving). Some commentators have over-looked this circular component of striving, and thereby mistakenly tried to explicate the rock's circular motion by means of the other two tendencies (e.g., Herival[23]).

Another possible source of confusion relates to the, for lack of a better term, "ontological" status of circular tendencies. As is evident above, Descartes' employment of a circular tendency seems to blatantly contradict his repeated claim that bodies only strive to move along straight-line paths. Some commentators have attempted to harmonize these divergent strands in the Cartesian story by relegating circular motion to a secondary, derived status; such that bodies really only strive to move rectilinearly, but that these tendencies are deflected by the intrinsic crowding of the plenum into a circular form. Garber appears to offer such a view in the following:

> At any instant, the shove that produces the motion in time can only be a shove in some one determinate direction. A succession of shoves can move the body in a curvilinear path, but any individual shove at a particular moment can push it in only one direction. It is in this sense that only rectilinear motion can be comprised in an instant, and in an instant only that required to produce rectilinear motion can be found. (Garber 1992, 286)

Apparently, Garber envisions Descartes' circular motion as brought about by instantaneous deflections from a rectilinear path. While this might be a successful interpretation of a Cartesian hypothesis of circular motion, it fails to do justice to Descartes' hypothesis of curvilinear *tendencies;* which, it is important to recall, are in effect at an instant ("Of course, no movement is accomplished in an instant; yet it is obvious that every moving body, at any given moment in the course of its movement, is inclined to continue that movement in some direction in a straight line, . . . (Pr II 39)." See section 3.2). Descartes specifically refers to the circular "tendency" (*tendere*) at an instant of its motion (i.e., the point A in Figure 11). Overall, it is thus not possible to pass off Cartesian circular tendencies as a species of motion, since motion, in contrast to tendency, is a durational phenomenon that occurs at the level of several instants (and not at an individual instant).

A devoted Cartesian could nevertheless invoke a variation on Garber's argument to overcome this difficulty. Rather than propose rotational motion as the derived phenomenon, a more successful strategy might simply allot this role to rotational tendencies. Accordingly, while moving bodies only *really* possess radially-directed tendencies, the ubiquitous plenum crowding induces a secondary, or derived, circular tendency (as required to form Descartes' large circling material bands). Freed from the confines of a plenum, bodies would never exhibit these additional, secondary tendencies; which thereby explains Descartes' repeated claim that bodies only tend to move along radial lines. This strategy, while resourceful, is unfortunately not supported by any textual evidence in the Cartesian cannon; but it is consistent with Descartes (apparently inconsistent) further claim that bodies have multiple tendencies towards motion (as above, Pr III 57). Therefore, unless more textual evidence were forthcoming, this proposed solution to the problem of circular motion must remain merely an intriguing possibility.

A basic question needs to be addressed at this point in our critical examination of Descartes' theory: Even if we do acknowledge an irreducible tendency towards circular motion in Cartesian bodies, in what sense does this constitute a form of circular inertia (as, for instance, endorsed by Beeckman's natural philosophy)? According to some commentators, the incriminating factor in Descartes' account of the rotating sling is his claim that the sling does not impede or affect the circular striving of the rock. R. Westfall, for instance, argues that "when he referred to the circular component as that part of the body's tendency 'the effect which is not impeded,' he returned unconsciously virtually to embrace the idea of a natural circular motion." (Westfall 1971, 81-82) A similar allegation is put forth by W. Shea: "the circular component of the motion of the stone is 'in no way impeded,' hence inertial."[24] In other words, if the body is not pressing (or being pressed by) the sides of the sling, then the body must be naturally "inclined" to move in a circle, thereby manifesting a form of circular inertia.

Yet, there is a general problem with this line of argument, for it ignores Descartes' additional hypotheses on the necessity of circular motion in a plenum setting (i.e., matter-filled universe). In Pr II 33, he states: "It has been shown . . . that all places are full of bodies . . . . From this it follows that no body can move except in a complete circle of matter or ring of bodies which all move at the same time." Descartes mandates these large-scale circular motions because there exist no empty spaces for a moving object to occupy. If the motion were not circular, the movement of a single body in his "indefinitely" large universe could produce a vacuum, or an indefinite material displacement that threatens the Cartesian conservation law (more will be said on this issue in later chapters). More importantly, for our purposes, is his stipulation that the circular motion of the bodies occurs synchronously "in such a way that [a body] drives another body out of the place which it enters, and that this body takes the place of yet another, and so on until the last, which enters the place left by the first one at the moment at which the first one leaves it." (Pr II 33) In the light of this additional hypothesis, the failure of the rotating rock to resist the sling does not thus constitute sufficient evidence for an instance of circular inertia in Descartes' plenum: since both the rock and sling partake in a large circular motion that occurs simultaneously, the rock *cannot* impede the sling. In short, there can be no resistance among bodies that move synchronously in the same direction. The instantaneous striving of the rock towards circular motion is matched by an identical striving of the sling, thereby accounting for the failure of the latter to impede the former. But, there exists no circular motion, and thus no instantaneous synchronized strivings, with respect to that portion of the sling contiguous with the outward radial direction of the rock's rotation. In that direction, the rock's striving is clearly impeded by a vast number of (relatively) stationary bodies (i.e., the bodies contiguous with the outer portion of the sling do not strive to move along the radial direction, so that the sling must "force" them into a large-scale circular motion of their own). Consequently, the appeal by both Westfall and Shea to the lack of circular impedance, as exhibited by the sling, does not of itself warrant an accusation of circular inertia in Descartes' natural philosophy.

There remains a problem with the Westfall/Shea critique, however, which apparently stems from an ambiguity intrinsic to Descartes' explanation of the rotating stone; namely, the fact that the philosophical implications of his theory (as regards circular inertia) lie somewhere in between both Galileo's and Beeckman's account of the phenomenon. Descartes appears to invoke, in a Beeckmanian fashion, a form of circular inertia in utilizing an unimpeded circular tendency; yet, in a Galilean manner, he presents the circular motion of the body as a composite, non-natural (non-inertial) state. Westfall refers to this sort of vague explanatory duality in the qualification that proceeds his claim that "[Descartes] returned unconsciously virtually to embrace the idea of a natural circular motion."

> I say 'virtually to embrace', for the conception of circular motion implicit in the analysis was closer to that held by Galileo, a conception in which circular motion appeared as a state of equilibrium between whatever constraint exists to turn bodies into a circular path and the force bodies exert against it. (Westfall 1971, 81-82)

If one is willing to admit this point, however (i.e., that Descartes' analysis is closer to Galileo's), in what sense can a charge of "circular inertia" be pinned on Cartesian natural philosophy? A mere tendency towards circular motion hardly counts as "(virtually) embracing" this notion, especially when it is recalled that; (i) the circular tendency is only one of several directed tendencies that the body possesses at an instant, and (ii) the body, if unimpeded by the sling, will never manifest a circular motion (but move along the tangent instead). In short, what the critics need to establish their case is an instance where Descartes explicitly employed a Beeckmanian form of unimpeded circular motion: that is, a motion which is "naturally" circular, and not constrained to follow a curvilinear path (or, alternatively, a "pure" tendency towards circular motion rather than just one of several such divergently directed instantaneous tendencies).

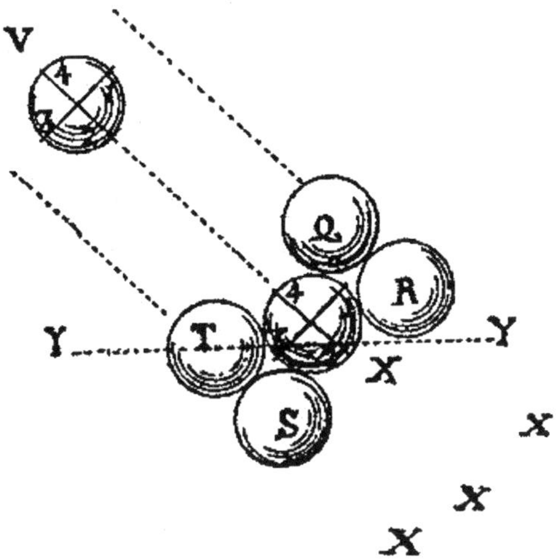

Figure 12. Descartes' illustration, from the *Meteorology,* of the rotational motion of air particles responsible for color. The particle V, moving obliquely from right to left, obtains a clockwise rotational motion upon striking the surface of the water YY.

Although largely overlooked by commentators, there is one piece of evidence within the Cartesian cannon which strongly endorses a Beeckmanian form of circular motion: viz., the letter from Descartes to Ciermans, 23 March 1638 (AT II 74). The context of the discussion is the Cartesian theory of the diffraction of light in a prism, first presented in the *Meteorology,* where Descartes ascribes a circular rotation to small spherical particles of air. (AT VI 330-331). Such particles constitute the means by which the pressure from a luminous source, i.e., light, is transmitted, while their different rotations account for the color spectrum (see Figure 12). In a previous correspondence (AT II 59-61), Ciermans had wondered why these small particles do not lose any of their rotational motion while traveling through space, to which Descartes responded:

> I do not see why it should seem to you that the particles of celestial matter do not maintain the rotational motion that gives rise to colors as well as the straight line motion of which light consists. We can equally well grasp both by our reasoning. I am convinced that, insofar as natural

events are concerned, we can think of nothing more accurate, namely that which better answers the rigor of mathematical computation. (AT II 74)

On Descartes' hypothesis, the passage of the particles through the prism alters their rotational speeds in such a manner as to account for the observed color spectrum of diffraction. Yet, if their motion is not disturbed, he sees no reason why the celestial particles "do not maintain [their] rotational motion" in the same way that they maintain "motion in a straight line". Mathematical considerations are also invoked as favoring this conclusion, although just what type of "mathematical computation" he has in mind is not disclosed. The primary importance of this passage is quite evident, nevertheless, for Descartes is assigning to rotational *motion,* which is usually conceived as non-inertial, the same metaphysical/ontological status as straight-line, inertial motion. In other words, celestial particles retain rotational motion in the same manner that they maintain straight-line motion, the latter being subsumed under his second law of nature. If Descartes' commentators are keen on locating evidence that reveals a Beeckman-type notion of circular inertia—such that he "virtually embraced" this notion—the letter to Ciermans must naturally form the center-piece of any such allegations, and not the ambiguous reference to a circular tendency as implied in the rotating sling examples.

Unfortunately, the Ciermans letter raises more problems than it solves; particularly, the question of how Descartes' diagnosis of the case of the rotating particles relates to his example of the rotating sling. In the latter, as described above, the stone exhibits multiple tendencies towards motion at any one instant, which explains why it would move along the circle's tangent if released. But, what tendencies (towards motion) does the rotating particle possess, if any? And if the particle's tendencies are, in fact, different from the stone's, why are they different (since all matter is essentially identical on Descartes' scheme)?

One possible resolution of this general puzzle might be found in the Cartesian theory of the three elements. The three basic elements are completely solid on Descartes' estimation, and thus have no parts (see, Pr III 46-53). The celestial particles utilized to explain the phenomenon of light he identifies as the secondary elements of matter, which are elsewhere described as "globules", and are "in proportion to there size, as solid as body can be, because they do not possess pores filled with other matter." (Pr III 123)[25] Without parts, it could be inferred that rotating particles only partake in the striving towards circular motion, and not both circular and radial strivings (which together give rise to the tangential), since there are no parts which could strive to flee the particle along a radial direction. (Or, alternatively, that any attempt by a "potential" part of this particle to flee along the radial direction would be perfectly balanced by the inversely directed striving of a "potential" part situated on the opposite side of the particle.) As for the rotating rock, on the other hand, the fact that the rock is only an individual member of the entire circle of moving bodies thus entails that it is free to strive to move along the radial path. This explanation, while consistent, faces various additional difficulties, nevertheless: first, there exists no known textual support for this hypothesis; and second, the entire ring of circling bodies would seem to qualify as a single body given Descartes' criterion for identity: "by *one body,* or *one part of matter,* I here understand everything which is simultaneously transported; even though this may be composed of many parts which have movements among themselves" (Pr II 25) If the complete circle of rotating objects does constitute a single body, then an identical argument could be made to demonstrate that it possesses no parts, and thus no radially directed strivings, as well.

3.4.3. Conclusion.

Overall, as regards Descartes' understanding of circular tendencies/motion, we can provide the following brief summary of the results of our investigation : first, while the various rotating rock examples admit a role for circular tendencies in Cartesian physics, it is the postulation of a natural circular motion (in the letter to Ciermans) that reveals the full extent of Descartes' use of circular inertial concepts; and second, this appeal to circular motion owes much more to a Beeckmanian, as opposed to a Galilean, interpretation of the phenomenon (which is contrary to the views of, at least, a few commentators). Only in the light of his discussion of the rotating celestial particles—a case long neglected by commentators—can one begin to grasp the true extent of Descartes' use of circular inertial concepts.

ENDNOTES

[1] One must be careful to distinguish the role of the Aristotelian "natural motions" in this synopsis: for example, the movements of the basic element "earth" towards the center of the universe, its "natural place," constitutes one of these natural motions. In short, natural motions do not require an outside or foreign cause (i.e., an imposed force), because they originate from some sort of internal principle intrinsic to the body. Once a terrestrial body has reached its natural place, however, it will remain in a state of rest unless moved by an external force, a "violent motion." Upon removal of this force, the body will once again seek its natural place.

[2] See, J. B. Barbour, *Absolute or Relative Motion?,* ibid., 196-203. I have also borrowed Barbour's terminology, here (i.e., "pusher"). In addition, it should be noted that Aristotle did not embrace an impetus theory as later proposed by the Scholastics. He sought the cause of the violent motions (as opposed to the natural motions) in an outside, external agent continuously in contact with the moving body, but not contained within the body (e.g., a hand, the air). See, R. Sorabji, *Matter, Space, and Motion* (Ithaca: Cornell University Press, 1988), 220-227. On impetus theory, see A. Maier, *On the*

Threshold of Exact Science, trans. by S. D. Sargent (Philadelphia: U. of Pennsylvania Press, 1982) chap. 4; and, J. E. Murdoch and E. D. Sylla, "The Science of Motion", in *Science in the Middle Ages,* ed. by D. C. Lindberg (Chicago: University of Chicago Press, 1978) 206-265.

[3] The apparent utilization of an "elasticity" type notion in Descartes' theory of internal "resistance" will also be discussed in chapter 4. See, G. W. Leibniz, "Critical Thoughts on the General Part of the Principles of Descartes," in *G. W. Leibniz: Philosophical Papers and Letters.* (Dordrecht: D. Reidel, 1969) 383-412.

[4] R. Descartes, *The Philosophical Writings of Descartes, Vol.3, The Correspondence,* eds. and trans. J. Cottingham, et al., (Cambridge: Cambridge University Press, 1991), 179.

[5] One must be cautious in attributing a composite structure to determinations, however. Although Descartes will allow a determination to be decomposed into constituent parts, he is not willing to ascribe to these parts an independent ontological status separate from the single motion of the body. In other words, the component parts of the determination of a single motion are not to be confused with the determinations of the several component parts of a single motion. Since a body's actual motion is not divisible, the component parts of its determination are only meaningful in relation to that one actual motion. For a lucid discussion of this distinction, see, P. Damerow, et al., *Exploring the Limits of Preclassical Mechanics.* (New York: Springer-Verlag, 1992) 119-120.

[6] R. Descartes, *Optics,* in *The Philosophical Writings of Descartes, Vol.1,* eds. and trans. J. Cottingham, R. Stoothoff, D. Murdoch (Cambridge: Cambridge University Press, 1984), 159. In this example, the ball travels more slowly after penetrating the sheet since it has transferred to it some of its quantity of motion.

[7] For a thorough discussion of the development of this problematic theory, see, D. Garber, *Descartes' Metaphysical Physics* (Chicago: University of Chicago Press, 1992), 234-248.

[8] For a thorough discussion of the philosophical importance of these concepts in the Early Modern period, see, R. Ariew and A. Gabbey, "The Scholastic background", in *The Cambridge History of Seventeenth Century Philosophy, Vol. 1* (Cambridge: Cambridge University Press, 1998), 425-454.

[9] R. Descartes, *Objections and Replies,* in *The Philosophical Writings of Descartes, Vol.2,* eds. and trans. J. Cottingham, R. Stoothoff, D. Murdoch (Cambridge: Cambridge University Press, 1984), 298.

[10] In actuality, the stone retains its position within the rotating sling due to a balance of the tangential centrifugal force and the centripetal pull of the hand towards the circle's center.

[11] Or, better yet, "center-fleeing" motions/tendencies, to avoid the philosophical implications of the modern understanding of the term "centrifugal".

[12] Westfall suggests that this aspect of Descartes' hypothesis is one of the last conceptual remnants of the Aristotelian/Scholastic theory of "natural circular motion" (i.e., a form of circular "inertia" that the Medievals commonly believed the motion of the planets to represent). R. Westfall, *The Concept of Force in Newton's Physics* (London: MacDonald, 1971), 82. Nonetheless, Descartes' theory only ascribes a single component of a body's "striving" to a circular path (at an instant). Given the combination of all the tendencies towards motion, the body will not move in a circular inertial motion if unconstrained (as demonstrated in the case of Descartes' stone upon release from the sling). See section 3.4.

[13] See, for example, R. Westfall 1971, 187-188.

[14] R. Descartes 1991, 131.

[15] See, for example, D. Garber 1992, 293-299; A. Gabbey, "Force and Inertia in the Seventeenth Century: Descartes and Newton." in *Descartes: Philosophy, Mathematics and Physics,* ed. Stephen Gaukroger (Sussex: Harvester Press, 1980) 230-320; G. Hatfield, "Force (God) in Descartes' Physics", *Studies in History and Philosophy of Science,* 10 (1979): 113-140; and M. Della Rocca, "'If a Body Meets a Body': Descartes on Body-Body Causation", in *New Essays on the Rationalists,* R. J. Gennaro and C. Huenemann, eds. (Oxford: Oxford University Press, 1999).

[16] It should be noted that Gabbey's interpretation of Cartesian force is rather complex and involves numerous additional postulates. For instance, Gabbey also understands force as a consequence of God's sustaining creative act (which grounds all existing things), and as a mode of body comparable to the closely related modes of "existence" and "duration." See, Gabbey 1980, 236-238.

[17] Besides their generally wrong predictions, there are numerous inconsistencies in Descartes' collision rules. On e of the first and best critiques of these rules belongs to Leibniz 1969, 383-412.

[18] That is, by declaring these states intrinsically or fundamentally "opposite" or "contrary" (as it can be interpreted from the French or Latin), Descartes reasons that motion and rest are mutually exclusive phenomenon that cannot transform or change into the one another when isolated from external influences. For a complete discussion of the role of the Scholastic logic of contraries in Descartes' natural philosophy, see, P. Damerow, et al., 1992, 82-91.

[19] See, e.g., M. Clagett, *The Science of Mechanics in Middle Ages* (Madison: University of Wisconsin Press, 1959) chap. 3-4; and, E. Grant, *Planets, Stars, & Orbs: The Medieval Cosmos, 1200-1687* (Cambridge: Cambridge University Press, 1994) chap. 18.

[20] For a nice discussion of the various influences on Descartes' natural philosophy, see, S. Gaukroger, *Descartes: An Intellectual Biography* (Oxford: Oxford University Press, 1995), and, W. R. Shea, *The Magic of Numbers and Motion: The Scientific Career of René Descartes* (Canton, Mass.: Science History Publications, 1991).

[21] Galileo Galilei, *Letters on the Sunspots,* trans. by S. Drake, in *Discoveries and Opinions of Galileo* (New York: Anchor, 1957) 113-114. For a recent discussion of these issues, see, W. Hooper, "Inertial Problems in Galileo's Preinertial Framework", in *The Cambridge Companion to Galileo,* P. Machamer, ed. (Cambridge: Cambridge University Press, 1998) 146-175.

[22] I. Beeckman, *Journal,* I, C. de Waard edition (The Hague: M. Nijhoff, 1939) 24.

[23] J. Herival, *The Background to Newton's 'Principia'* (Oxford: Oxford University Press, 1965) 47, 54. Westfall also points out this misreading of Descartes' theory; 1971, 93-94, fn. 49.

[24] W. R. Shea 1991, 281-282.

[25] The concept of "solidity" in Cartesian natural philosophy is quite complex. See, chapter 4 for an attempt to clarify this, and related, notions.

CHAPTER 4

MATTER AND SUBSTANCE IN THE CARTESIAN UNIVERSE

The overall goal of our investigation is to explore possible formulations of a Cartesian physics that can circumvent some of its well-known internal inconsistencies, such as those raised in Newton's famous argument against Descartes' theory of space and motion. However, in order to investigate the prospects of successfully devising such a response, we will need to explore further the details of Descartes' laws of nature. While chapter 5 will investigate the origins and function of the Cartesian conservation law for "quantity of motion," the present chapter will examine the conditions or stipulations that Descartes appended to his collision rules, as well as the specific character and content of Descartes' hypotheses of matter and (material) substance. Up to this point, it has been largely assumed that Descartes' laws, and his corresponding concept of matter, will present no special difficulties when transferred to a plenum setting (i.e., his matter-filled world). Yet, as will be shown, this marriage presents a number of obstacles that threaten to undermine Descartes' entire project. I will argue, nevertheless, that many portions of Descartes' major scientific treatise, the *Principles of Philosophy,* demonstrate that he was fully aware of, and capable of addressing, many of the difficulties brought to light by the conjunction of these seemingly incompatible partners, i.e., his quasi-absolutist natural laws and a plenum environment.

4.1. "Perfect Solidity" and The Natural Laws

Among the ideal conditions that appear in his *Principles of Philosophy* (Pr II 37-52), Descartes remarks that his seven rules on the impact of material bodies "could easily be calculated . . . if [the two colliding bodies] were perfectly solid (*durus*). . . (Pr II 45)." Although this stipulation was intended to assist in the application of the collision rules (as instances of his third natural law), the complex and puzzling concept of "perfect solidity," or "perfect hardness," is perhaps also notable for having generated several divergent interpretations among recent commentators. In some instances, "perfect solidity" has been translated into the modern dynamical locution "perfect elasticity," which denotes a class of material bodies that return to their original shape, volume, etc., after deforming under impact.[1] The translators of the first complete English edition of the *Principles,* V. R. Miller and R. P. Miller, for example, assume that Descartes had elasticity in mind in the passage quoted above.[2] While it is true that most commentators regard Cartesian bodies as "inelastically hard" (i.e., they do not alter their shape during collision), even a few of these scholars have been unable to completely resolve their doubts. For instance, R. S. Woolhouse has recently concluded that "since the 'before and after' of a perfectly hard collision is the same as the 'before and after' of a perfectly elastic one, there is some justification for the sometimes-made assumption that by 'hard' [solid] Descartes *really* means 'elastic'."[3] Notwithstanding the merits of this elastic/inelastic controversy, it would appear that such disputes overlook a more fundamental question that lies at the heart of the "perfect solidity" issue: Does Descartes' use of the "perfect solidity" concept encompass only the interactive, *collision* properties of bodies (i.e., how they behave under impact), or are other individual, non-interactive factors implicated as well, such as their internal composition and configuration of elementary particles? Contrary to the prevailing viewpoint (or lack there of) among commentators, we will present a case for the latter interpretation of perfect solidity, claiming that it constitutes the only means of correlating much of the information found in the latter portions of the *Principles* with the natural laws put forth in Pr II (Part II of the *Principles*). Despite its apparent connotations, "perfect solidity" pertains to the internal constitution of the basic particles that make up macroscopic bodies, and only indirectly the dynamic properties manifest in bodily collisions.

In order to demonstrate that Cartesian solidity is not merely a stipulation on bodily collisions, however, we will need to explore the interrelationship between the Cartesian conserved property of quantity of motion (often described as "size times speed") and the three bodily properties of volume, surface area, and quantity of matter. The specific function of these latter three properties is an issue that has received scant attention among Cartesian scholars, but it is crucial to a full understanding of Descartes' physics. For example, in the canonical presentation of the conservation law for the quantity of motion, Descartes' *Principles* specifically incorporates bodily surface area as a key factor in the determination of quantity of motion:

> We must however notice carefully at this time in what the force of each body to act against another or resist the action of that other consists: namely, in the single fact that each thing strives, as far as is in its power, to remain in the same state . . . . One which is at rest has some force to remain at rest, and consequently to resist everything which can change it; while a moving body has some force to continue its motion, i.e., to continue to move at the same speed and in the same direction. This force must be measured not only by the size of the body in which it is, and by the [area of the] surface which separates this body from those around it; but also by the speed and nature of its movement, and by the different ways in which bodies come in contact with one another (Pr II 43).

Thus, the quantity conserved in the motion and impact of bodies, which Descartes refers to as "quantity of motion," is determined by three factors: size, surface area, and speed (where speed is conceived as a non-directional scalar property, see chapter 3). Although the role of surface area is not revealed at this stage in the *Principles,* the derivation of his natural laws in Pr II apparently equates a body's size with its volume, and its quantity of motion with the product of its size and speed (e.g., "when one part of matter moves twice as fast as another twice as large, there is as much [quantity of] motion in the smaller as in the larger," Pr II 36). Given his thesis that spatial extension (in three dimensions) constitutes the essential property of material substance (Pr II 11), it is probably not surprising that the alleged identification of Descartes' term “size” with “volume” is often accepted as an elementary fact of Cartesian science.[4] Nevertheless, it is also true that Descartes' conservation law implicates both size *and* surface area in the determination of quantity of motion, two properties whose exact interrelationships, and hence contribution to the conservation law, are never clearly detailed in the Cartesian theory of matter.

The problem of harmonizing Descartes' sporadic references to surface area, volume, and quantity of matter with his principal use of the term 'size' has long been a source of irritation among Cartesian scholars. Many who are aware of the ambiguous contribution of Cartesian matter to the natural laws have only mentioned this difficulty in passing, since their main research concerns others aspects of Descartes' physics. To illustrate, in an important essay on Cartesian force, M. Gueroult's only reference to these issues is the observation that "the notion of mass identified with volume remains very obscure in Descartes."[5] In other cases, scholars will draw attention to the imprecise meaning of Descartes' term by placing it in quotation marks (i.e., “size”), as does A. Gabbey in the following: "for Descartes, . . . the force of motion of a body, . . . is the product 'size' × speed . . . ."[6] Overall, the sentiment of many of these Cartesian researchers is probably best captured by D. M. Clarke's remark that "the details of [Descartes'] theory are never sufficiently developed, so that one finds the same rather vague references to size, surface area, resisting media, and speed . . . ."[7]

Nevertheless, the details of Descartes' theory are less vague than most commentators have assumed. To demonstrate this point, sections 4.1.1 through 4.1.4 will examine how the bodily properties of volume, surface area, and quantity of matter are integrated by Descartes' concept of perfect solidity. On the basis of this discussion, we will then return, in section 4.1.5, to the question posed at the outset: i.e., Is "perfect solidity" chiefly a collision, or composition, property of macroscopic bodies?

4.1.1. Rarefaction and Condensation.

Descartes' definition of solidity first appears in Pr II of the *Principles,* Articles 54: "Those bodies whose particles are all contiguous and at [relative] rest, are solid." In this passage, "contiguous" is the term requiring further elaboration, since the appeal to the "relative rest" of the particles pertains to Descartes' rejection of a binding force among the infinitely-divisible particles of matter (Pr II 20, 55). Essentially, Descartes treats the observed phenomenon of varying bodily density, or, as he phrases it, "solidity," through an appeal to the spaces between the particles of matter. With respect to those processes which either decrease or increase the size of material bodies (labeled, respectively, rarefaction and condensation), he states: "rarefied bodies are those with many spaces between their parts which are filled by other bodies. And rarefied bodies only become denser when their parts, by approaching one another, either diminish or completely eliminate these spaces; . . . (Pr II 6).” Thus, bodies whose particles are contiguous (i.e., they are not separated by an influx of foreign matter) are deemed "solid" (see Figure 13). Descartes evidently found these natural processes of varying density rather disturbing, for they "might lead one to doubt whether the true nature of body consists in extension alone," a remark that also explains their presence at so early a stage in the *Principles*. Yet, only in Pr III 48-52, are we first introduced to the hierarchy of material elements responsible for the swelling and shrinking of these large macroscopic bodies. Briefly, Descartes procures a threefold subdivision of matter in order to explicate the underlying mechanisms that operate his matter-filled, or plenum, world. These basic particles, largely differentiated by size and function, are: (i) the large, macroscopic third elements of matter, and (ii) the much smaller, globule-shaped second elements of matter; while the minute debris formed from the collisions of the second and third elements, known as (iii) the first elements of matter, serve to fill the lacunae manifest between these larger particles (see Figure 14).

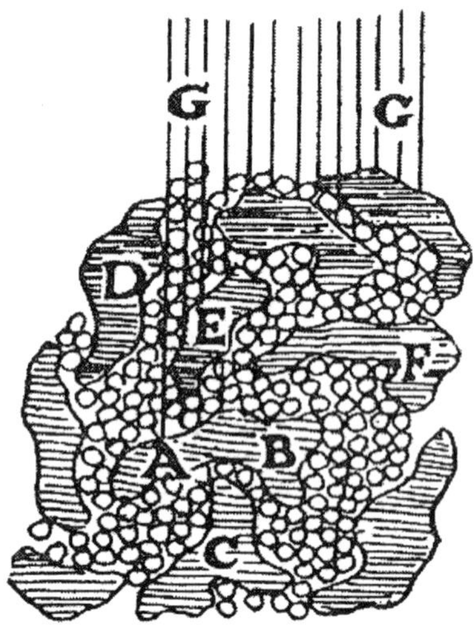

Figure 13. A illustration from the *Principles* that reveals the many smaller particles inside the pores of larger bodies.

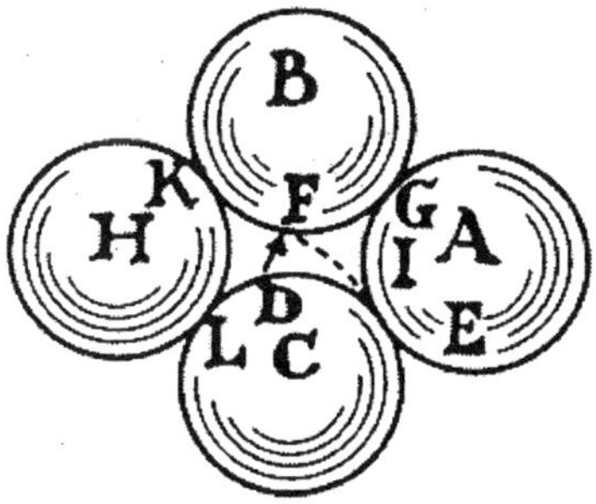

Figure 14. The constantly moving first elements of matter fill in the gaps between the larger elements, such as the triangular space FGI between the secondary globules A, B, C, and H (AT III 92).

In Pr III of the *Principles,* Descartes presents a somewhat more elaborate analysis of the problem of solidity. Quite possibly, he felt compelled to furnish a systematic explanation of this phenomenon after reflecting upon the variety of diverse behavior produced by identically-sized bodies in resisting or sustaining motion—behavior which today we would call inertial effects. That is, there often exists a disparity among bodies of the same spatial volume, such as two identically sized globes composed, respectively, of gold and wood, in resisting changes to their states of motion (e.g., one is much harder to move than the other). One of the principal motivating factors in the formulation of the Cartesian theory of solidity is the need to explicate the origin of these "inertial" effects. In fact, a discussion of the motions of celestial bodies occasions Descartes' next attempt at a definition: "the solidity of [a] star is the quantity of the matter of the third element, . . . in proportion to its volume and surface area (Pr III 121)." As defined, solidity is thus a function of three variables: quantity of third element matter, surface area, and volume. Since the distinction between these three quantities, and their role in affecting density, is often misunderstood, we shall examine this three-part interrelationship below.

4.1.2. Volume, Quantity of Matter, and the Agitation Force.

At one point in the examination of solidity, Descartes utilizes his ratio of quantities to resolve the problem, just described, of divergent motions that originate from bodies of equal volume. He explains:

Thus, here on earth, we see that, once moved, gold, lead, or other metals retain more agitation, or force to continue in their movement, than do pieces of wood or rocks of the same size and shape; and consequently metals are also thought to be more solid, or to contain more matter of the third element and smaller pores filled with the matter of the first and second elements (Pr III 122).

Descartes' remarks contain an implicit conjecture on the origins of inertial effects (although, as previously noted, Descartes did not hold the modern concept of inertia):[8] A body's "force to continue in its movement," or "agitation," is directly proportional to its amount of third element matter. Therefore, provided two bodies of equal volume (and equal speed), the more solid object will possess the greater quantity of third element matter, and consequently produce a greater tendency to continue in its motion (or agitation). This interpretation of the passage is verified by his discussion of the motions of stars: "The force which [a star] acquires from its motion . . . to continue {to be thus transported or} to thus move, which I call agitation; must be estimated neither by the size of its surface area nor by the total quantity of matter which composes it, but only by the quantity of the third element matter . . . (Fr Pr III 121)." In Pr II, furthermore, a moving body's "quantity of motion" (or "size times speed") constitutes a measurement of its "force to continue its motion, i.e., to continue to move at the same speed in the same direction (Pr II 43);" which is the same description Descartes provides for his force of agitation in Pr III (see section 4.1.1). Quantity of motion is hence a gauge of agitation force, a conclusion that will later assume importance.[9]

At this juncture, one may begin to question the overall consistency of combining Descartes' agitation force hypothesis with his theory of matter. If, as Descartes believes, matter is mere spatial extension, then why should the agitation force of, say, a body entirely composed of tertiary matter differ from that of a body (of equal volume) containing only first element matter? Since all matter is extension, it would seem that both bodies should behave in exactly the same manner. Descartes, however, is quick to provide a rationale for this association. He reckons that, because the individual motions of a collection of elementary particles are not entirely unified, a volume of secondary globules cannot produce an agitation force equal to that of an identical volume of tertiary matter. In an insightful passage concerning stellar motion, he compares the agitation force of a star composed of third element matter against the force produced by an equal volume of secondary globules:

> Because these globules are separated from one another and have various individual movements; although their united force acts against the star, they cannot all unite their force simultaneously in such a way [as to ensure] that no part of their force is wasted. In contrast, all the matter of the third element . . . forms one single mass which is moved together as a whole, and thus all the force which it has to continue in its motion is applied in a single direction (Fr Pr III 124).

Put simply, the variably-directed motions of the individual globules of secondary matter (and, presumably, first element matter) lessen the total agitation force of the composite volume *in any single direction*. The motion of the star is not subject to these same effects, on the other hand, since all of the matter of third element constitutes a unified whole which moves in a single direction. Therefore, it is not the case that tertiary matter possesses some internal property that makes it qualitatively different from the other material elements; rather, it is simply the relative rest of the particles that comprise the third element which account for its role in determining the agitation force of *macroscopic* bodies, such as stars. Since the three types of

Cartesian matter are identical in all respects except their relative size, Descartes' agitation theory is thus also applicable to particles entirely composed of primary and secondary matter. For instance, among secondary particles, the globules that possess more secondary matter will harbor the greater agitation forces (as will be evident in later sections).

Although Descartes' reasoning is rather ingenious, it does not entirely justify his correlation of a macroscopic body's agitation force with its quantity of third element matter. For, even if the individual motions of the first and second elements cannot simultaneously unite their forces in a single direction "in such a way [as to ensure] that no part of their force is wasted," it is still the case that they will contribute some force, albeit small, in that given direction. The force generated by these particles will probably be insignificant in comparison to the force provided by the tertiary matter, but it is still a distinct force, and thus it must make some contribution to the overall agitation force of the composite body (i.e., the star). In fact, Descartes seems to admit this interpretation of his theory: "As for the matter of the first or even the second element, it is continually leaving [a] star and being replaced by new matter. Consequently, this new matter approaching cannot retain the force of agitation acquired by the matter which has already left, which, in any case, was very small (Pr III 121)." If, as Descartes concedes, the departing first and second elements have acquired a force of agitation from the star, then this acquired force must have contributed something to the star's total quantity of motion, a force these particles obtained while inside the star. On this rendering of the evidence, it thus appears that a Cartesian body's agitation force should be equated with its total quantity of matter, and not just its quantity of third element matter.

It is unfortunate that Descartes did not adequately explain his reasons for disregarding the contribution of the primary and secondary elements to the agitation force of macroscopic bodies, but his brief statements may indicate two possible motives for this decision. First, as specifically mentioned in the quotation above, the primary and secondary elements of matter are continuously leaving a star and being replaced by new matter. Hence, these minute particles are best seen as foreign bodies that only temporarily occupy the macroscopic host body, and do not assist in the composition of that body. Of course, if these foreign particles do not qualify as constituent members, then the effects of their individual agitation forces on the host body can be ignored. This interpretation has the added bonus of nicely correlating with a claim made early in Pr II: "whatever extension there is in the spaces between [a body's] parts must in no way be attributed to it, but to whatever other bodies fill those spaces (Pr II 6)." Nevertheless, it should be noted that this form of response does not sit well with Descartes' definition of an individual body; where "by *one body,* or *one part of matter,* I here understand everything which is simultaneously transported; even though this may be composed of many parts which have movements among themselves (Pr II 25)." As for the second possible rationale, since Descartes holds that the agitation force acquired by the first and second elements is "very small," he may believe that this minute quantity can be conveniently ignored when determining the body's overall agitation force. In other words, one can secure a fairly accurate approximation of a body's quantity of motion by simply taking the product of its total quantity of tertiary matter and speed. Regardless of which construal of Descartes' reasoning we adopt, his correlation of agitation force and third element matter in the case of macroscopic bodies remains somewhat obscure; but it will assume great importance below (in section 4.2.2) when we discuss the impact of such composite bodies.

4.1.3. Surface Area and the Agitation Force.

Besides quantity of tertiary matter, a body's agitation force is also substantially influenced by the magnitude of its surface area. On the whole, Descartes is well aware that the agitation force of a body can be modified by simply changing its shape. For instance, a golden sphere can assume shapes that will allow a less-dense wooden sphere to possess a "greater agitation; . . . if [the golden sphere] is drawn out into threads or forged into thin plates or hollowed out with numerous holes like a sponge, or if it in any other way acquires more surface area, in proportion to its matter and volume, than the wooden sphere (Fr Pr III 122)." In this case, the magnitude of an object's surface area is clearly implicated in the resulting agitation force: The larger the proportion of surface area to third element matter, the smaller the resulting force. This formula likewise holds for the individual globules of secondary matter. In discussing the agitation force of various sized globules, he argues that "the smaller [globules] have less force, because they have more surface area in proportion to the quantity of matter . . . than the larger ones . . . (Pr III 125)."

Yet, the relationship between surface area and quantity of matter in Cartesian natural philosophy is a rather complex, and possibly uncertain, affair. As quoted above, Descartes claims that a star's agitation force is a sole function of its quantity of third element matter, with surface area playing no role. In the very next article, though, he openly admits that a body's surface area can greatly change the magnitude of its agitation force. Contradictions of this sort bedevil much of the Cartesian theory of solidity and agitation, which provides a possible explanation for their lack of serious coordinated analysis. In essence, Descartes seems to desire a simple correlation between a body's agitation force and its quantity of second or third element matter; but, he also recognizes the important role of surface area in modifying this force, a variable he cannot completely ignore.

Descartes' uncertainty regarding the role of surface area is a dilemma that can be traced back to the definition of quantity of motion in Pr II. On the whole, Descartes was cognizant of the plenum's capacity to retard the motions of bodies: "It is obvious, moreover, that [bodies] are always gradually slowed down, either by the air itself or by some other fluid body through which they are moving. . . (Pr II 38)." If we conjoin this observation with Descartes' comments on the motions of various shaped material objects (as noted above), then it seems plausible to infer that Cartesian surface area merely functions to *change* a body's existing agitation force, rather than assist in *constituting* that force. More carefully, a large surface area "slows down" a moving body by increasing the number of plenum particles it encounters along its path: the greater the magnitude of the body's surface area, the more particles it will confront (as opposed to a smaller shape), and hence the more quantity of motion it will lose or transfer to the surrounding plenum.[10] This interpretation of the role of surface area in Descartes' physics has been duly noted by R. S. Westfall: "Descartes asserted frequently that the quantity of surface on which other bodies can impinge modifies the force of a body to continue its motion (1971, 70)." This realization most likely prompted Descartes to incorporate surface area into his definition of the quantity of motion, the force that is conserved in all bodily collisions (as quoted in section 4.1.1). "This force must be measured not only by the size of the body in which it is, and by the [area of the] surface which separates this body from those around it; but also by the speed and nature of its movement . . . (Pr II 43)."

Nevertheless, the prospects of "quantifying over" the retarding effects of surface area must have presented a serious obstacle to the formulation of the Cartesian conservation law. Prior to the analysis of the collision rules, and *just after* his definition of quantity of motion, Descartes strives to eliminate this extra variable by insisting that his "[colliding bodies are] separated from all others both solid and fluid in such a way that their movements would be neither impeded nor aided by any other surrounding bodies . . . (Fr Pr II 45)." Accordingly, with the aid of this additional idealized condition, it is no longer necessary to take into account a body's particular shape when calculating its quantity of motion. In the presentation of the collision rules, the mitigating effect of surface area on the agitation force has been ruled out by definition, thus explaining its conspicuous absence in the derivation of these seven hypotheses.[11] Yet, Descartes' ideal conditions only prevail for the collisions depicted in this section of the *Principles*. In Pr III, the reemergence of surface area as a factor in the motions of bodies clearly indicates that such conditions are no longer in effect. It is a general mistake, therefore, to disregard the influence of surface area when ascertaining a body's quantity of motion outside the context of the idealized conditions utilized in the collision rules.

4.1.4. Agitation and Solidity: Towards a Synthesis.

In determining the factors involved in Descartes' agitation theory, we have examined thus far the functional relationship between surface area and quantity of matter, and between volume and quantity of matter; but the exact means by which all three quantities are integrated into a single concept or formula remains largely unexplained. Fortunately, in a discussion of the mitigating effects of surface area on a star's agitation force (relative to the agitation force of secondary globules), Descartes provides an outline of this three-part interrelationship:

> It can happen that [a star] has less solidity, or less ability to continue its movement, than the globules of the second element which surround it. . . . For these globules, in proportion to their size, are as solid as any body can be, because we understand that they contain no pores filled with other . . . matter; and because their figure is spherical; the sphere being the figure which has the least surface area in proportion to its volume. . . (Pr III 123).

In this two-part analysis, Descartes essentially provides the clearest formulation of his theory of solidity, and of the means by which a body's agitation force is linked to its solidity. With respect to his first claim, a globule completely packed with (secondary) matter is more solid than any other globule of identical size; where, as exercised in this quotation, the notoriously obscure term "size" apparently denotes volume. That is, without pores filled with matter (presumably first element), these globules are fully condensed (as defined above), and thus possess the highest degree of solidity. Given the earlier reference to the solidity of stars, it is safe to assume that this account of solidity must also hold for bodies entirely composed of tertiary matter, rather than just the globules composed of secondary matter. As a result, we can translate the expression "contiguous" in Descartes' first definition of solidity with the degree of bodily porosity. For the second part of his definition, Descartes claims that a spherical body is more solid than any other *figure* of the same "size" (volume), since the sphere manifests the smallest proportion of surface area to volume.[12] Presumably, these identically-sized objects possess a similar quantity of second or third element matter; for, if they did not, a highly rarefied spherical body could conceivably retain more solidity than a fully condensed non-spherical body of similar volume, in direct violation of the first part of the definition. We can generalize this section of Descartes' hypothesis as follows: given two bodies of identical volume and identical quantities of third element matter (or second, if it is a globule), the body possessing the smallest surface area will harbor the greatest agitation force. Therefore, inasmuch as agitation force is linked to solidity, our three quantities—volume, surface area, and quantity of matter—are essential ingredients in the magnitude of this force.

All told, at least one important lesson can be extracted from Descartes' complex and troublesome theory of solidity: Any attempt to simply identify a body's quantity of motion with its volume and speed, as seemingly implicated in the definition of quantity of motion, is inconsistent with the analysis of solidity offered in Pr III. Given our analysis, it is thus evident that, *if used outside of the context of the idealized collision rules,* quantity of motion tacitly expresses an intricate relationship between a body's volume, surface area, and its total quantity of second or third element matter (besides speed). In various circumstances, these variable magnitudes determine the agitation force, and hence quantity of motion, of all physical bodies in the Cartesian plenum. In other words, inasmuch as agitation or quantity of motion are directly dependent upon our three quantities, the theory of solidity presented in Pr III informs and governs the operation of the conservation law under the normal, non-idealized conditions that prevail in the Cartesian plenum. Most expositions of Descartes' laws of nature do not disclose or investigate this important aspect of the *Principles,* but it is crucial to a full understanding of Cartesian dynamics.[13] In fact, one can find in the Cartesian literature numerous attempts to isolate spatial volume, or quantity of matter, as the sole contribution of Cartesian matter to the conservation law (size times speed).[14] Yet, these readings of quantity of motion only hold for the highly idealized conditions assumed in the collision rules; where, as discussed in section 4.1.3, the role of surface area in modifying the agitation force has been negated by Descartes' exclusion of disrupting plenum effects.

4.1.5. Perfect Solidity and the Natural Laws: A Proposal.

We can now return to the analysis of the impact behavior of Descartes' "perfectly solid" bodies, which has been the motivating cause of our lengthy investigation. As we have seen, Descartes couples the agitation force to an intricate relationship among three different bodily quantities. Provided this theory, it would seem an almost impossible task to secure a systematic *quantitative* description of the inertial tendencies and conserved motions of material bodies. In order to produce such a law, one would need to determine the exact means by which the relative proportions of quantity of matter, volume, and surface area, contribute to the overall conserved motions of the colliding system. Yet, rather than undertake these potentially unrealizable determinations, Descartes circumvents the problem by (1) simply confining the scope of his collision laws to the impact of completely solid bodies, and (2) ignoring the plenum's capacity to retard bodily motion via bodily surface area. More precisely, if the globule or body is fully condensed (contains no pores), then it embodies as much second or third element matter (respectively) as its volume permits. No longer is it necessary to compute the ratios of quantity of matter to total bodily volume among the colliding bodies—under this requirement, all that is obligated is a measurement of their relative bodily volumes. In addition, without the need to consider surface area, a body's total quantity of second or third element matter, or volume (given perfect solidity), can now be conveniently equated with its agitation force (after taking its product with speed, of course). It is thus no longer necessary to calculate how different shapes affect the quantity of motion of bodies with identical volume.

As utilized in the collision rules (Pr II 46-52), the requirement for perfect solidity can be therefore largely viewed as a stipulation for completely dense, pore-less bodies. That is, provided the evidence of the entire *Principles,* Descartes' appeal to perfect solidity partially amounts to a restriction on the potential ratio of a colliding body's quantity of matter to its total volume. The motivation underlying this restriction stems from the obvious need for a simplified and manageable treatment of the variables influencing a body's agitation force, and hence quantity of motion. As mentioned at the outset, despite the strong dynamic connotations, perfect solidity is *not* a requirement exclusively allied with the impact behavior of material bodies—other non-interactive properties, such as internal composition, form an important part of this concept.

Put slightly differently, one can view perfect solidity as "an attempt to isolate behavior that can be regarded as fundamental given Descartes' metaphysics of matter," as J. Carriero has suggested.[15] Since Cartesian matter is pure extension, the perfect solidity criterion frees the collision rules from the internal complications—via elementary particles—that beset the interactions of most macroscopic bodies (see section 4.1.2). Without the need to factor in these interfering effects, perfect solidity thus allows the collision rules to describe the "pure" or "actual" collision behavior of Descartes' extended bodies. However, even on this interpretation (which is complimentary to the one advanced here), perfect solidity still remains primarily a stipulation on bodily composition, and not a description of how bodies *behave* during impact (e.g., by losing and regaining their original shape, or completely reversing their direction of motion). Of course, there is a sense in which any specification of the composition of bodies is relevant to their collision properties, due to the simple fact that a body's constitution affects its interactions. But, bodily composition does not directly outline or detail such interaction behavior: rather, bodily composition, i.e., perfect solidity, is only indirectly concerned with the outcomes of bodily impact. In the next section, the analysis of the concept of "rigidity" will further substantiate this rendition of Descartes' perfect solidity concept.

4.2. Non-Idealized Conditions and the Natural Laws.

This section examines the prospects of successfully applying the Cartesian conservation principle without relying on the highly idealized conditions presupposed in the collision rules; that is, without limiting the scope of the natural laws to the interactions of just poreless bodies. As will become evident, many facets of Descartes' theory of matter and motion can generate substantial difficulties for this undertaking. Nevertheless, we will consider various methods of utilizing the insights gained from our study of Cartesian solidity, and of our forthcoming analysis of Cartesian rigidity, to overcome these obstacles.

4.2.1. The Problem of Size Invariance.

Besides definitional or computational simplicity, there may exist an aspect of Descartes' solidity hypothesis that specifically concerns the effects of bodily motion and impact. To demonstrate this point, we need to investigate the origins of Descartes' plenum universe. According to the Cartesian cosmological hypothesis, all space (matter) was initially divided into homogeneous parts of equal size, and impelled with a conserved quantity of motion: "God, in the beginning, divided all the matter of which He formed the visible world into parts as equal as possible and of medium size . . . . [Also] He endowed them collectively with exactly that amount of motion which is still in the world at present (Pr III 46)." Eventually, the collisions of these equally-sized spatial parts formed the three Cartesian elements of matter, as well as the vast diversity of material bodies comprised from these elements. The initial impact of Cartesian matter could not have been elastic, consequently, because of the absence of the fragmented particles necessary for the composition of porous bodies. Without pores, the original bodies in Descartes' universe were perfectly solid. Yet, these bodies were clearly not perfectly solid in the sense required for Descartes' collision rules, since their impact ultimately resulted in a loss of size through fragmentation. When objects disintegrate or shed particles in this manner, their total quantity of motion will inevitably decrease.[16] More specifically, macroscopic bodies that do not possess determinate volumes do not possess determinate quantities of motion, which is defined as the product of speed and size (or volume, recalling the interdependence of these concepts as discussed above). Of course, a Cartesian will probably insist that any lost quantity of observable bodily motion is merely transferred to the realm of the surrounding microscopic particles, thus preserving the total universal quantity of motion. This form of response, although possibly correct, does not remedy the plight of Descartes' collision rules, however. If the Cartesian collision rules are to be applied successfully *at the level* of macroscopic objects, it is necessary that they conserve the total *bodily* quantity of motion by maintaining an invariant magnitude of bodily size. Hence, besides the absence of bodily pores, Descartes' perfect solidity criterion may also be interpreted as sanctioning a property of unchanging bodily size (or volume).

Moreover, the union of the Cartesian theory of matter with a plenum universe poses major obstacles for the successful application of the Cartesian conservation law. This conflict probably stems from Descartes' denial of a material binding force; for when this theory, that relatively resting particles constitute solid bodies, is conjoined with any explication of bodily motion in a plenum, it becomes very difficult to accommodate the further contention that moving objects do not change volume over time. If the particles that comprise a solid object are at rest (relative to one another), the force exerted by the surrounding bodies and particles during motion, let alone impact, would seem quite sufficient to dislodge large numbers of them. Accordingly, over a given temporal interval it may be impossible to posit a determinate volume for any moving Cartesian body. Yet, solid bodies that persistently shed minute particles as they travel and collide will experience a decrease in their quantity of motion with their corresponding reduction in volume, once again presenting great difficulties for any attempt to apply the Cartesian law to the collisions of macroscopic bodies over any temporal span.

Evidence may exist, furthermore, to support the contention that the lack of a material binding force influenced Descartes' view of impact. In the illustrations of colliding bodies that accompany the letter to Clerselier, dated 17 February 1645, it is potentially significant that the cubes in one of Descartes' pictures possess equally sized rectangular surfaces on their sides of collision: i.e., the collision or contact surfaces of the two cubes are congruent (see Figure 15).[17] Provided the Cartesian denial of binding forces, this congruence may amount to a logical or practical consideration, for there would seem no means of preventing larger cubes from breaking apart (on contact with smaller cubes) without equal contact surfaces.[18] For example, if the collision surface of a cube C extends beyond that of a second cube B, many particles on the periphery of C's surface will not encounter any opposing B particles upon impact, and thus continue their motion past the contact surface (resulting in C's disintegration). That the prospects for such incidents may have troubled Descartes is also disclosed in his letter to Clerselier. In picturing the collision of two unequally sized bodies, he merely increases the length of one of the cubes while preserving the congruence of their contact surfaces. If this interpretation of Descartes' illustrations is correct, then we can add a further stipulation to the ideal condition for perfect solidity; namely, that two bodies manifest identical impact surfaces.

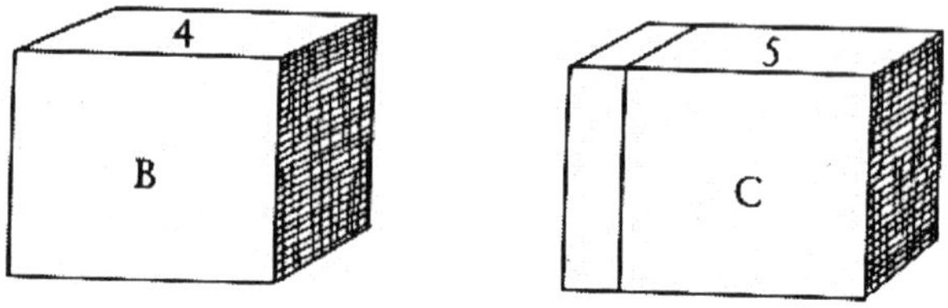

Figure 15. Descartes' depiction of two unequally sized bodies in his letter to Clerselier. Note that both B and C are situated so as to collide upon sides possessing congruent surfaces.

This construal of the Clerselier illustrations faces problems, nonetheless. With respect to these same collisions (among objects of different size), a second drawing included in the Clerselier letter does not display identical contact surfaces, nor do any of illustrations contained in the *Principles* (see, Pr II 45-52—see Figure 16). In addition, it is not clear just how congruent contact surfaces can prevent the disintegration of colliding Cartesian bodies. As previously discussed, if relatively resting particles are prone to separate when confronted by an external material agent, then the forces exerted by the numerous plenum particles would seem quite sufficient to disperse a Cartesian body even when it is not impacting. In essence, congruent contact surfaces will not compensate for the lack of a material binding force.

Figure 16. One of Descartes' illustrations of colliding bodies in the *Principles,* AT II 46-52. Note that the collision surfaces are not congruent.

4.2.2. Rigidity.

In order to further examine the application of Descartes' conservation law to the impact of plenum bodies, it will be necessary to explore the Cartesian concept of "rigidity," a notion that essentially constitutes a theory of elasticity. Towards the end of Pr IV (on terrestrial phenomena), he states:

> Glass is rigid: that is to say, it can be somewhat bent by external force without breaking but afterwards springs back violently and reassumes its former figure, like a bow. . . . And the property of springing back in this way generally exists in all hard bodies whose particles are joined together by immediate contact rather than by the entwining of tiny branches. For, since they have innumerable pores through which some matter is constantly being moved . . . , and since the shapes of these

pores are suited to offering free passage to this matter . . . , such bodies cannot be bent without the shapes of these pores being somewhat altered. As a result, the particles of matter accustomed to passing through these pores find there paths less convenient than usual and push vigorously against the walls of these pores in order to restore them to their former figure (Pr IV 132).

On Descartes' estimation, a "rigid" body is capable of returning to its original configuration after impact due to the action of matter, presumably first element, contained within its pores. These primary elements of matter recover the body's initial shape by pressing against the walls of the pores during the contraction of impact. Cartesian rigidity, the analogue of elasticity, thus stands in sharp contrast to the Cartesian definition of solidity: the former concept invokes pores or channels within the structure of material bodies, while the latter notion requires fully condensed bodies completely devoid of such conduits. To verify this material classification, one needs only to recall Descartes' synopsis of the completely solid secondary elements: "these globules, in proportion to their size, are as solid as any body can be, because we understand that they contain no pores filled with other . . . matter (Pr III 123)." Without pores, Cartesian bodies are thus incapable of changing and regaining their bodily shape under impact. Needless to say, these comments make it abundantly clear that, contrary to the suggestions of some commentators, Descartes did not intend perfect solidity to mean perfect elasticity.

In addition, by confidently asserting that rigidity "generally exists in all hard bodies," Descartes' exposition on elastic phenomena makes it clear that most physical bodies are not perfectly solid. One must exercise caution in interpreting this claim, however, since the standard explications of the terms "hard" and "rigid" somewhat overlap: Descartes is not claiming that all bodies, including the perfectly solid ones, are elastic (rigid); rather, he is merely pointing out the non-trivial fact that most *seemingly* perfect solid bodies are actually elastic. On a deeper level, one may in fact read into his statement a denial of the very existence of perfectly solid bodies. Descartes' observations thus reflect and corroborate the tacit assumption that perfect solidity is an *ideal condition* imposed on the domain of his conservation laws. Furthermore, as an historical aside, Descartes' analysis of rigidity belies the simplistic judgment that all Cartesian bodies are inelastically hard. On the contrary, his comments reveal an intuitive awareness of the fundamental elasticity of most, if not all, macroscopic objects; a conclusion closely akin to the later elastic theories of Leibniz and John Bernoulli.[19]

Returning to the problem of implementing the conservation law, it is important to note that the utilization of Descartes' hypothesis concerning tertiary matter and the agitation force, first introduced in section 4.1.2, greatly assists in conserving the quantity of motion of colliding "rigid" bodies. As mentioned, rigid bodies possess pores or channels filled with elementary particles of matter, particles whose presence or absence occasions the phenomena of rarefaction and condensation. When these elastic bodies collide, consequently, it would seem that many of the small foreign particles housed in the objects, i.e., the primary and secondary matter, must be expelled or ejected during the contraction phase of the impact (when the distortion of the body during the brief instants after contact compresses its pores). Because of the Cartesian identification of matter and space, only by emitting matter can the body reduce the volume of space it occupies. Likewise, the primary and secondary elements of matter will somehow need to filter back into the object after the contortion phase of the impact to "puff" it back up to its original size. The exact manner by which this process takes place is decidedly unclear given Descartes' brief comments on the problem of elasticity—presumably, some sort of "hinge" mechanism on the surface of porous bodies could be invoked to meter the flow of particles both in and out of the channels. Nevertheless, if we adopt Descartes' correlation of tertiary matter and agitation force, then one aspect of "rigid" collisions is evidently clear: the reduction and increase in a body's overall volume during the temporal period spanned by an elastic collision will not vary its total quantity of motion, since all the particles ejected and recovered will not be third element matter. Descartes' stipulation of an "agitation" force, a theory that seemed somewhat unmotivated when first presented (in section 4.1.2), thus proves invaluable in applying the conservation law to the collisions of the normal, non-idealized bodies that predominate in the Cartesian plenum (i.e., rigid bodies).

Before proceeding to the next section, we should offer some concluding remarks on the success of Descartes' theory of matter as implicitly contained within the idealized collision rules. With the benefit of scientific hindsight, Descartes' decision to formulate his laws of motion upon the groundwork of his particular theory of matter was rather unfortunate. His conservation principle, which demands a firm material theory to operate effectively, is severely handicapped by the quirks of his physical program; most notably, by the interference of the plenum and the lack of a material binding force. In short, it seems impossible to integrate all of the Cartesian theories surveyed in this essay into a coherent system of dynamics. Yet, although the solidity thesis harbors various inconsistencies, it exhibits a striking awareness of the diverse factors involved in the inertial motion of bodies, as well as offering a sophisticated attempt at integrating these disparate elements into a single manageable formula. Unfortunately, the Cartesian notion material "substance," the topic we shall pick up in the remaining section of this chapter, possesses very few of the advantages of Descartes' theory of solidity, but many of its obstacles.

4.3. Cartesian Corporeal Substance

If the duality of Cartesian mental and material substance has long been a source of contention among scholars, the precise nature of Descartes's notion of material substance *per se* has proven to be equally troubling. The difficulties, hinted at in previous chapters, mainly center upon the nature of individual corporeal substance, with its associated constellation of space-based concepts, such as place. Put simply, if a substance is defined as an independently existing entity, but a single part of extended substance requires the existence of its contiguous neighbors (given the matter-filled plenum) to delimit its surface, then in what sense is an individual corporeal part of the plenum *really independent,* and thus a substance? It would seem that all corporeal entities, or parts of extension/space, necessarily require the existence of all other corporeal individuals (or rest of extension/space). This conclusion, which seems to undermine the very possibility of an individual material substance, was much debated among the Cartesians (e.g., Cordemoy, Rohault, Spinoza),[20] and is still as essential topic in any investigation of Descartes' natural philosophy (e.g., Rodis-Lewis, Garber, Des Chene).[21]

Many commentators, finding the argument presented above rather persuasive, have reasoned that only the whole of Descartes' extended matter achieves the requisite independent status to qualify as a substance (such as, Keeling, Woolhouse, Anderson, to name only a few).[22] In this section, it will be argued that, contrary to the received opinion, a fairly plausible case can be made for individual corporeal substances in Descartes' plenum. The evidence for this conclusion is not only based on the specific details, and implications, of Descartes' hypotheses, but also on his actual use of substance terminology as regards material objects.

4.3.1. The Problem of Individual Corporeal Substance.

Throughout his written work, Descartes often referred to individual material objects, such as a person's clothing, as substances: "a man who is dressed can be considered as compounded of man and clothes . . . . although clothes are substances (AT VIII-2)." There are many other passages presenting a similar view: e.g., "the two halves of a portion of matter, no matter how small they may be, [are] two complete substances (AT III 477; see also, Pr II 11, 55, 60; AT VII 44, 222)." As for the definition of substance, in the *Principles of Philosophy,* he asserts that "by 'substance' we can understand nothing other than a thing whose existence is such that it needs no other thing in order to exist (Pr I 51)."[23] This work also presents his well-known thesis that extension constitutes the essential property, or nature, of material substance ("extension in length, breadth, and depth constitutes the nature of corporeal substance . . . ."; Pr I 53). As is often the case with Descartes' physical hypotheses, the exact nature of the relationship between extension and material substance is shrouded in uncertainty. Indeed, the language that Descartes uses to describe the relationship would seem to suggest that they are identical; i.e., that there is only a conceptual, and not real, distinction between the two (Pr II 8-12).[24] The independence of extended substance from mental substance is, on the other hand, one aspect of Descartes' theory upon which all commentators find agreement (see Pr I 54).

For our purposes, the problem with corporeal substance stems directly from Descartes' ideas on place, which also implicates his notions of body and motion: "*the transfer of one piece of matter or of one body, from the neighborhood of those bodies immediately contiguous to it and considered at rest, into the neighborhood of others.* By *one body,* or *one part of matter,* I here understand everything which is simultaneously transported; . . . (Pr II 25)." Thus, an individual "body" is defined by its containing, or common, surface with its contiguous neighboring bodies; where "by 'surface' we do not understand any part of the surrounding bodies, but only the boundary between the surrounding and surrounded bodies, . . . (Pr II 15; Descartes also deems this common surface, 'external place')." The conceptual difficulties inherent in Cartesian space and motion are legion (as should be quite evident to the reader by now), yet these hypotheses raise relevant concerns about Descartes' penchant for referring to individual corporeal objects, like clothes, as substances. If, as noted above, a substance can exist independently of any other thing, then the direct dependence of a Cartesian body on its contiguous neighbors (to delimit its surface) apparently undermines its claim to substancehood. For later commentators (notably Spinoza and Leibniz, as quoted below), Descartes' analysis of the possibility of a vacuum, particularly a hypothetical scenario wherein God removed all the matter from within a vessel, became the focal point of these worries about the ontological status of individual material bodies:

> If someone were to inquire as to what would occur if God removed the whole body contained in a vessel and did not allow anything to take the place of the body that had been removed, the answer must be that the sides of vessel would thus become contiguous to one another. For, when there is nothing between two bodies, they must necessarily touch one another; . . . (Pr II 18).

On Descartes' estimation, a vacuum would be an empty space, or extension without matter, and is thus a contradiction in terms. So, removing the matter within a vessel is the same as removing its extension, with the inevitable outcome that the sides would now be contiguous.[25]

While some Cartesians dutifully followed Descartes' analysis of God-induced vacua, albeit with minor modifications, a number of natural philosophers found the implications of such hypothetical scenarios so unsettling that they were led to repudiate outright the very idea of an individual material substance. The thesis that advocates a non-individual corporeal substance, also known as "monism", is forcefully presented by Spinoza in the *Ethics*:

> If corporeal substance could be so divided that its parts could be really distinct, why could not one part be annihilated, the rest remaining, as before, connected with one another? . . . For of things which are really distinct the one from the other, one can be and remain in its own position without the other. Since, therefore, it is supposed that there is no vacuum in Nature . . . , but that all parts must be united so that no vacuum can exist, it follows that they cannot be really separated, that is to say, that corporeal substance, . . . cannot be so divided (Part I, Prop. 15).[26]

The reference to the "really distinct" parts of matter (presumably) pertains to an argument put forth in Descartes' *Principles;* namely, that since the extended plenum can be divided by our minds into separate parts "clearly and distinctly," these parts must be really separate substances (Pr I 60; this passage will be discussed at length below). Spinoza reasons that, like the sides of the vessel, such parts are not really distinct or separated since they depend on their contiguous neighbors to retain their position (and thus lack the necessary independence to qualify as individual substances). Similar sentiments are voiced by Leibniz, who concludes: "In what is truly one substance there are not many substances."[27] Many present-day commentators have also favored the monist interpretation of Descartes' complicated material hypotheses. For instance, S. V. Keeling concluded that "a chair, a table, and a stone are not really, hence cannot be clearly understood to be, three separate extended *substances*. . . . They *really* are divisions within a single homogenous substance which is existent space (145)."

Most monistic theorists remain silent or, at the least, sketchy on the ontological status of individual material objects in their one extended substance. Presumably, Descartes' doctrine of bodily "modes", such as shape or motion, play the principal individuating role of in the homogenous plenum (where modes are the particular manifestations of bodily extension: e.g., the same volume of extension having either the mode of a circle or square; Pr I 64-65). For instance, M. Gueroult, one the chief proponents of this view, argues that among the variety of different modes of extension there are "modes properly speaking which alone merit this name in all respects, such as shapes or bodies, ..."[28] If we conjoin the definition of "body" (and motion) in the *Principles* with the monist theory, the picture that begins to emerge is of a single, undivided plenum that nevertheless manifests a variety of migrating, individual geometric shapes (=bodies), much like the many ripples that move across a pond's surface. A similar view is often attributed to Spinoza, who is probably foremost among the Cartesians in advocating a material substance monism. For instance, on J. Bennett's reading of Spinoza, individual bodies are regions of extension/space endowed with body-like attributes: "If there is . . . a pebble in a region R, what makes this true is the fact that R is *pebbly,* where 'pebbly' stands for a certain monadic property that a spatial region can have."[29]

T. Lennon, taking a more Malebranchian line, confines the individuating properties of bodies to the phenomenal realm, thus denying that there is any real motion or change in Descartes' one corporeal extension. Where Bennett conceives motion as neighboring spatial regions successively possessing a bodily property, Lennon maintains that "motion ontologically is the flow of sensations", much like "a shape distinguished from the visual field by a difference in color will move when qualitatively the same shape is distinguished in an adjacent area of the field by the same color."[30] Despite its ingenuity, Lennon's Draconian interpretation of Cartesian bodily motion, to use his own appraisal (201), is conspicuously incompatible with Descartes' numerous references to motion as a mode of extension, and not, presumably, as a manner of perceiving extension. On this new reading, motion would be rendered more a mode of mind than body, apparently contradicting Descartes' insistence that "figure cannot be understood except in an extended thing, nor can motion, except in an extended space; nor can imagination, sensation, or will, except in a thinking substance (Pr I 53)." There is no textual evidence to support the claim that the mode of motion, as mentioned above, should be interpreted any different from the modes of sensation, willing, etc.; for they are both modes *in* their respective substances (i.e., motion in matter, and sensations in mind). (We will return to these general issues in chapter 6.)

Although the scarcity of ontological and epistemological details associated with the monist theory remains an obstacle, the real difficulty with this view would seem to lie in its lack of textual support; for Descartes makes no overt references to individual material bodies as modes of a single extended substance. (In fact, one gets the distinct impression that most commentators are simply interpreting Cartesian metaphysics from a Spinozistic perspective.) There is one notable exception to this general lack of evidence, however, and it is often invoked by monists to confirm their view. This passage, put forth in the Synopsis to the *Meditations,* is worth quoting in full:

> Body, taken in the general sense, is a substance, ... But the human body, in so far as it differs from other bodies, is simply made up of a configuration of limbs and other accidents of this sort; whereas the human mind is not made up of any accidents in this way, but is a pure substance. For even if all the accidents of the mind change, so that it has different objects of the understanding and different desires and sensations, it does not on that account become a different mind; whereas a human body loses its identity merely as a result of a change in the shape of some of its parts.[31]

In commenting upon this passage, R. S. Woolhouse states:

> The idea then seems to be that unlike human minds, human bodies (as also stones and horses) are not numerically different individual substances, but pieces of corporeal substance or body as such, in the way that pieces of lead are just that—pieces of lead, and not 'leads' [i.e., not different individual lead substances] (22).

Moreover, Woolhouse insists that the discussion (quoted above) from the *Meditations* "should be taken as Descartes' considered view [of corporeal substance] (23)."

There are various difficulties with Woolhouse's analysis, however. First, in order to accept his interpretation of the above passage, one must render the expression, "body, taken in the general sense," into the phrase "body, taken as the entire plenum (or whole of extension)," since only this translation would seem to secure the desired monist outcome. Not only does this interpretation of Descartes' wording strain credibility, but he must also deny that the intended meaning was the more plausible phrase, "body taken as any part of the plenum (or possibly the entire plenum)", or, more simply, "body taken *generally.*" A truly faithful treatment of the term "generally" should allow for "any part of, or all extension", rather than the more limited "all of extension". Second, Woolhouse's version of this quotation appears incompatible with Descartes' further explication of the ontological status of the human body and mind, mainly as regards their substancehood and identity. What Descartes seems to be stressing in this passage is that the nature of the human body, as a substance or unique entity, depends upon a distinct correlation of its numerous parts; a correlation of parts, moreover, that is not required for the substancehood of mind or of *any body* not possessing a complex internal structure—that is, not required of body taken *generally*. A human body, unlike any of its parts, is an interconnected whole whose very nature, or identity, depends on an elaborate structural arrangement of its parts (so that if you were to remove or rearrange these various constituent parts, it would no longer be the same body). In fact, the Cartesian hypothesis referred to by Woolhouse not only fails to undermine the substancehood of individual corporeal bodies, but Descartes' further elaboration of this very issue (in the *Objections and Replies*) would seem to favor the cause of the anti-monists. For example:

> Anyone who states that a man's arm is a substance which is really distinct from the rest of his body does not thus deny that the arm belongs to the nature of the whole man. And saying that the arm belongs to the nature of the whole man does not bring about the suspicion that it cannot subsist in its own right (AT VII 228; see also AT VII 222).

Before turning to a defense of the non-monist version of corporeal substance, a few remarks are in order as regards the "modal" interpretation of Cartesian bodies, such as Gueroult's theory (mentioned above). Quite simply, the most telling objection to this line of thought is its straightforward incompatibility with the extant discussions of Cartesian substance. In the *Principles,* which contains the most elaborate discussion of substances and their modes, all the specific examples of corporeal substance involve individual bodies (e.g., stones), and not the whole of extension: "If a stone is in motion and is square-shaped, I can understand the shape of the square apart from the motion and, oppositely, the motion without the shape of the square; but I cannot understand either the motion or the shape apart from the stone's substance (Pr I 61)." The modal thesis likewise runs afoul of a section in the *Sixth Replies,* where Descartes insists that "the surface [of a body] is merely a mode and hence cannot be part of a body (AT VII 433)." That is, given his understanding of a surface as the boundary between the contained and containing bodies (Pr II 15, see above), Descartes feels confident to proclaim that "a body is a substance, and a mode cannot be a part of substance (AT VII 433)." Since a boundary does not form a part of a substance, it qualifies as a mode—unfortunately, the modes that comprise Gueroult's theory would not seem to satisfy this same criterion: if bodies are parts of matter (Pr II 25, as above), and bodies are also modes of the one corporeal substance, as the modal thesis advocates, then modes (=bodies) are parts of a substance.[32]

4.3.2. Countering the Monist Interpretation.

The monist case rests on the assumption that a substance, being an independent entity, cannot depend on another "thing" for its existence. On one reading of Spinoza's argument, it is the position (place) of the body relative to its neighbors that violates this, as we shall call it, "independence" clause. Yet, does this form of dependence, i.e., positional, really constitute a violation of Descartes' criterion of substancehood? If a part of the plenum were removed by God, thus resulting in a different arrangement of the surviving plenum inhabitants, it is true that the contingent status of the original positions of these parts would have been clearly demonstrated, but it is not at all obvious that their existence would have likewise been rendered dependent. (In addition, it should be noted that all Cartesian substance is dependent upon God; Pr I 51) If one were to apply Spinoza's reasoning scrupulously in all cases, it would seem to follow that *all substances* would fail the independence test. For example, although the mind is a separate substance from the body, it depends on the body's sense organs for a large portion of its mental content; viz., sensory ideas of color, texture, taste, etc.. Note the symmetry in these two dependency cases: just as a given body depends on its neighbors for its bodily property of position, so the mind depends on its sense organs for its mental sense ideas. Consequently, there would seem to be a defect in Spinoza's critique of individual corporeal substance: while some of the properties of a corporeal substance may be dependent, such as place, this does not necessarily undermine the object's independence as a distinct entity.

A monist would likely respond to this interpretation of Spinoza's argument by insisting that: (i) the intended target was not the property of position per se, but the very extent or boundary of an individual corporeal substance; which (ii) is a problem exclusively confined to material, and not mental, substance. In other words, the independence clause is violated since an individual material substance relies upon its contiguous neighborhood to provide its identity (via the containing surface). Without these contiguous neighbors, there would be no individual body, or plenum part. The same does not hold true for mental substance, however. Even if God were to eliminate all but one mind, the resulting state-of-affairs would not thereby invalidate the existence of the remaining mind. This is surely a persuasive interpretation of the monist critique (and possibly more what Leibniz had in mind), but notice that the emphasis would seem to have shifted away from the original "vessel" example. Rather than a "hole" in extension, as with the vessel thought experiment,[33] a more troubling prospect for the non-monist would appear to be the opposite hypothetical case; i.e., a single material substance surrounded by an empty, non-extended space. Without contiguous neighbors, what would provide the means of delimiting, or "marking off", the extension of the individual substance? Is an individual corporeal substance—which is essentially finite, bounded extension—even a coherent notion under these circumstances? This line of monistic criticism would seem to draw additional support from Descartes' rejection of the vacuum. In the *Principles*, he reasons that there can be no limit to extended substance, "for wherever we imagine those limits to be, we can always imagine not merely other indefinitely extended spaces beyond them, but also clearly understand that they are as we conceive them. . . (Pr II 21)."

Nevertheless, even granting the strength of the monist position, the argument against individual corporeal substance does not necessarily go through. The monist case, as reconstructed above, relies on two key assumptions: (1) that an individual substance is delimited or defined by its contiguous neighbors (in order to secure the common surface); and (2) that space is necessarily "indefinitely" extended. From these Cartesian doctrines, the monist correctly concludes that not only is it impossible to remove any part of extension, but that such a removal would undermine the "contiguous neighborhood" means of delimiting individual substance. Yet, to decisively refute the non-monist, they need to prove an additional claim; namely, (3) that the lack of a contiguous neighborhood would eliminate or nullify the very existence of the extended material part previously bound by that neighborhood. Appealing to the factual impossibility of a vacuum cannot greatly assist the monist cause in pondering this hypothetical scenario, moreover. The main question concerns the following hypothetical, and not factual, case: If the contiguous neighborhood were annihilated, would it also nullify the existence of the remaining portion of matter (previously contained within its common surface)? Now, it is not at all clear that this question can be provided an affirmative answer, let alone if it even has an answer—but that is the response, i.e., (3), that the monist must conclusively prove in order to refute the non-monist opponent. Put differently, while the contiguous neighborhood certainly defines the surface or extent of the contained material part, one cannot conclude from this form of dependency that the plenum part would cease to exist if that neighborhood were removed.

On the monist's behalf, it could be argued that a portion of matter without a containing surface (external place) is simply incoherent given the definitional parameters of Descartes' natural philosophy. Hence, all talk of existence is futile if a neighborless plenum part cannot be conceived. This objection carries a great deal of plausibility, especially provided Descartes' denial of the vacuum and his notion of clear and distinct ideas (Pr I 30). The non-monists have a few defensive strategies they can fall back upon, however. First, if the monist thesis ultimately amounts to the claim that "existence" cannot be predicated of a neighborless extended substance, then the existence of the *entire plenum* would be correspondingly threatened. Since it is "indefinitely" extended, the plenum *cannot* be enclosed within a contiguous neighborhood of containing bodies; and thus the monist would have to conclude that it, too, cannot exist.[34] Now, the monist might contest this embarrassing outcome by invoking the fallacy of composition: in particular, it is not always true that what is required of the parts of a whole, or each external place, is also required of the whole, i.e., the plenum comprised of all external places. Whether or not the need for an external place transfers from the parts to the whole is an open question, of course, but it does raise doubts about the general plausibility of a monist position that would deny existence to a neighborless, uncontained material substance.

The second reply to the alleged conceptual impossibility of a neighborless corporeal substance is based on an important passage in the *Principles*; a passage, furthermore, that constitutes the most powerful evidence in support of the non-monist. As noted above, Descartes stipulates that the parts of an extended substance which are separable in thought are really distinct, and thus are substances in their own right:

> A Real [distinction] exists properly only between two or more substances: and we perceive these to be really distinct from each other from the single fact that we can clearly and distinctly understand one without the other. Since, knowing God, we know with certainty that he can accomplish whatever we understand distinctly. For instance, from the single fact that we have an idea of extended or corporeal substance . . . , we know with certainty that it can exist; and that if it does exist,

each part of it which is capable of being delimited by our minds is really distinct from the other parts of the same substance. (Pr I 60)

Of crucial importance, here, is Descartes' claim that if we can distinctly understand "one [thing] without the other", then they really are distinct, independently existing substances. A part of extended substance can be distinctly separated, or delimited, from other parts in this same manner. Thus, Descartes affirms that each material part can be distinctly understood "without the other" parts, a conclusion which closely matches the "neighborless" interpretation of individual corporeal substance put forth above.

As a "real" distinction, moreover, the individuality of the plenum parts is grasped by the mind a priori. On this issue, D. Garber makes the important observation that the discussion of individual substance, in Pr I 60, is somewhat separate from the problem of individuation in the later *Principles* (176). As he explains, the key element in Cartesian physics, and not metaphysics, is motion. In Pr II (*Principles,* Part II), as described above, a body or part of matter is definitionally bound to the concept of motion; yet motion has little, if anything, to do with the type of thought-based individuation of substance prevalent in Pr I. Garber's point concurs with our line of investigation, therefore, since the problem of the "empty vase", as well as the possibility of a neighborless individual substance (i.e., lacking a common surface), are conceptual, thought-based problems that do not involve motion (see, endnote 25). Unfortunately, Descartes occasionally confuses the issue by employing the term "body", which implicates motion, in his analysis of the empty vessel, which we have diagnosed as a purely space/plenum conceptual problem (not implicating motion). These confusions are probably inevitable, however, given the mere "conceptual distinction" between Cartesian space and body.

Of course, even if individual substances are delimited by thought alone, this does not forego the need for a contiguous neighborhood (or common surface, external place, etc.): it only means that the common surface separating an individual substance from the rest of extended substance is, likewise, a conceptual construct. Given the homogeneous expanse of extension, a person can imagine any number of "distinct" partitions of this substance into individual, independent elements (as will be discussed at greater length below). The conceptual nature of substance, as presented in Pr I 60, thereby makes the delimitation of individual substances relative to the conceiving individual, as Garber also notes (176). E. Grosholz finds this conceptual method of demarcation mysterious,[35] but it is no more mysterious than, say, the Cartesian causal interaction of mind and body, two different substances without a common essence (Pr I 53)—consequently, the unintuitive connotations of a particular interpretation of Cartesian doctrine may not necessarily count against its accuracy in representing Descartes' intended view.

Returning to the monist challenge, it would seem that the ability to conceive distinctly separate portions of matter, via Pr I 60, thus corresponds to our ability to conceive distinctly a plenum without limit. Yet, how can both of these statements be simultaneously true (namely, really independent parts but a real whole)? There is only one interpretive option that would seem to remain faithful to Descartes' (apparently contradictory) doctrines; viz., that if God were to annihilate all but one part of extension, the existence of that one remaining part would not be likewise annulled. The surviving plenum part would lack the contiguous neighborhood to materially delimit its surface, yet its existence could still be conceptually understood, or "delimited". This point is not easy to grasp, but an analogy with the concept of space, which is identical to extension and matter for Descartes (Pr II 9-12), may be of some assistance. Although space is conceived as a unity, with all its parts existing at once, the integrity of a single extended part of space would not appear to be compromised if one could (quite miraculously) annul the rest of space. Admittedly, the process by which space could be so removed is difficult to conceive, even granting the many historical precedents for just such a view.[36] But, the very existence of the surviving portion of that space, and not necessarily its border or surface, is quite a bit easier to comprehend.[37]

Put differently, distinctly conceiving the existence of an individual corporeal substance is not the same thing as distinctly conceiving its surface or boundary within the totality of extended substance—and this may be the conceptual distinction that Descartes was trying to make in simultaneously advancing both vacuum-less extension and individual corporeal substance. Some commentators have openly, and quite understandably, conflated these two issues. For instance, in critiquing Descartes' corporeal substance hypotheses, M. Grene states, "if we want to count bodies, they must be clearly and distinctly separable . . . . So where is the 'distinctness'?"[38] She appears to equate "distinctness" with the ability to "count them [i.e., corporeal substances], separating each by the space it occupies . . . (101)." Yet, Descartes is explicit in tying "distinctness" to our mental ability to "understand one without the other", which is not necessarily a process that differentiates bodies by *embedding* them in space. As argued above, if the annihilation of the rest of the plenum does not nullify the existence of the one surviving part, then the existence of that part has indeed passed the "one without the other" test of substancehood. This method of accounting for individual substance, moreover, does not appeal to a background spatial framework or a contiguous neighborhood to achieve its goal. It merely relies on the hypothetical case of an extension-less space, or vacuum. The fact that a vacuum is, in reality, an impossible state-of-affairs for Descartes does not detract from its use as a means of conceptually determining substancehood.

The problem for this construal of individual material substance is that it lacks support from subsidiary discussions of substancehood in the Cartesian corpus. Pr I 60 is apparently the only detailed explanation that would confirm the interpretation advanced above. Nevertheless, the ensuing portion of that article compares the individuality of corporeal substances with the individuality of mental substances:

> And, similarly, solely because every one understands himself as a thinking thing, and can by thought exclude himself from all other substance, whether thinking or extended; is certain that each one of us is really distinguished from every other thinking substance and extended substance (Pr I 60).

This passage provides crucial supporting evidence for the interpretation of Descartes' phrase "one without the other" assumed above, since thinking substance can, in fact, exist without material substance (as is well known from the *Meditations*). That is, just as one portion of matter can be understood to be distinct from any other portion, so can a thinking substance be excluded, to use Descartes' term, from any other substance. Descartes would appear to be using such terms and phrases as, "distinct", "exclude", and "one without the other" interchangeably in these passages. And, as argued above, this seems to confirm the view that "individual" substances are ascertained by a conceptual process of excluding, removing, etc., all other things in order to determine if the original thing can still be coherently conceived. Attempting to imagine a neighborless plenum part may be difficult, but it may be no more difficult than trying to grasp dualistic causation; and many Cartesian commentators have been able to entertain this latter notion without undue apprehension.

In determining the more fundamental conceptual factors that may have prompted this view of material substance, suspicion would tend to fall upon the Cartesian notion of a “surface”; which, as explained above, is a common boundary that belongs neither to the contained or containing bodies. Its quasi-independent status is confirmed in the following excerpt from Descartes' explanation of the Eucharist:

> The intermediate surface between the air and the bread is not different in reality from the surface of the bread, or from the surface of the air which touches the bread; in fact the three surfaces are a single thing that differ only in our thought. That is, if we call it the surface of the bread, we mean that even though the air which surrounds the bread has changed, the surface remains always numerically the same, given there is no change in the bread, . . . (AT IV 164).

In general, the common surface can remain the same despite a change in either the contained or containing bodies. For our purposes, the more important realization is that a body's surface is invariant under a change of containing bodies; that is, a different set of containing bodies can replace the original containing bodies without altering the common surface. Based on this insight, it is possible that Descartes may have drawn the inference—a rather problematic inference—that one could still meaningfully refer to a body and its surface without the existence of *any* containing bodies. Of course, the one remaining body would not possess a common surface in the literal sense (since there are no longer any containing bodies), but it would possess a sort of vestigial conceptual possibility of a common surface (or, pursuing a mathematical metaphor, it is the limiting case of a common surface as the existence of the containing bodies goes to zero).

In conclusion, even if the explication of Descartes' natural philosophy advanced above remains somewhat troublesome, conceptually speaking, it has at least one other clear advantage over the monist interpretation: namely, it strives to accommodate the Cartesian doctrine of vacuum-less space *and* Descartes' numerous references to individual corporeal substance, which is a source of textual support not open to the monist. The monists have tried to render Descartes' natural philosophy palatable to their metaphysical intuitions. Correspondingly, they have interpreted the mutual interdependence of plenum parts as a violation of the tenet of Cartesian substance. Yet, as we have seen, the monist's understanding of these issues is not Descartes', and thus the exact nature of his complex definitional distinctions warrants an analysis more faithful to the Cartesian texts.

4.4. Concluding Remarks.

With the benefit of historical hindsight, Descartes' decision to formulate his laws of inertial motion within the setting of an Aristotelian/Scholastic plenum was rather unfortunate, as was his corresponding decision to adopt problematic notions of matter and material substance. His quasi-mathematical descriptions of straight line motion and conserved interaction forces, which seemingly plead for "space" to operate effectively, are constrained both physically and figuratively by his matter-filled universe and the quirks of his bodily concepts. Overall, this problem is symptomatic of relational theories which eschew reference to non-material aspects or structures of the spacetime arena. Inevitably, they are faced with the task of explaining (or "explaining away") the inertial motions and forces observed in the behavior of material objects while confined to an ontological domain devoid of everything except the physical bodies. His attempts to overcome these and other difficulties, such as the need to delineate plenum bodies or explain the agitation force of various sized objects, suffer a similar fate: basically, it seems impossible to integrate all of the Cartesian theories surveyed in this chapter into a coherent system of dynamics. Yet, this conclusion should not be taken to invalidate or lessen the overall value of Cartesian natural philosophy; since the very influence of Descartes' work on the succeeding generations of scientists is enough to dispel this simplistic notion. The ideal condition of "perfect solidity," one of the main subjects of this chapter, is a case in point: although the solidity thesis harbors various inconsistencies, it exhibits a striking awareness of the diverse factors involved in the inertial motion of bodies, as well as offering a sophisticated attempt at integrating these disparate elements into a single manageable formula.

ENDNOTES

[1] "The property of recovery of an original size and shape is the property that is termed *elasticity*." A. E. H. Love, *A Treatise on the Mathematical Theory of Elasticity* (New York: Dover, 1944), 92. However, the term 'elastic' receives various interpretations. E. J. Aiton reasons, for example, that a body is elastic only if it rebounds in the opposite direction (while presumably recovering its initial shape) "Descartes term 'hard' . . . cannot be equated with 'elastic,' since the hard bodies sometimes rebound and sometimes move together." E. J. Aiton, *The Vortex Theory of Planetary*

Motions (MacDonald: London, 1972), 39. Consequently, Aiton interprets "perfect solidity" as a dynamic, or collision, property of bodies.

[2] See, Miller and Miller, in Descartes, 1983, 64, fn. 44.

[3] He continues: "There is good reason for Descartes to have considered perfectly hard [solid], perfectly elastic bodies, for when soft and perfectly inelastic bodies collide the 'quantity of motion' is not conserved." R. S. Woolhouse, *Descartes, Spinoza, Leibniz: The Concept of Substance in Seventeenth Century Metaphysics* (London: Routledge, 1993), p. 114. For a recent example of an "inelastically hard" reading, see, D. Garber, *Descartes' Metaphysical Physics* (Chicago: Chicago University Press, 1992), 357.

[4] Rohault's influential Cartesian text of the late seventeenth century provides a nice example of the contribution of bodily "volume" to quantity of motion: "If a body of *two Cubic Feet* runs through a Line *sixty Foot long,* it has twice as much Motion, as a Body of one *Cubic Foot,* which runs through the same Line . . . ." J. Rohault, *A System of Natural Philosophy, vol. 1* (written 1671), trans. J. Clarke and S. Clarke (1723), (New York: Johnson Reprint Corp., 1969), 43.

[5] M. Gueroult, "The Metaphysics and Physics of Force in Descartes," in *Descartes: Philosophy, Mathematics, and Physics,* ed. S. Gaukroger (Sussex: Harvester Press, 1980), 228, fn. 100.

[6] A. Gabbey, "Force and Inertia in the Seventeenth Century: Descartes and Newton," in *Descartes: Philosophy, Mathematics, and Physics,* ed. S. Gaukroger (Sussex: Harvester Press, 1980), 245.

[7] D. M. Clarke, *Descartes' Philosophy of Science* (Manchester: Manchester University Press, 1982), 217. Clarke is certainly aware of the function of perfect solidity, however, as I will later comment. Another excellent discussion of the complexity of the factors influencing Descartes' physics as regards collisions is P. McLaughlin, "Force, Determination, Impact", in *Descartes' Natural Philosophy,* ed. by S. Gaukroger, S. Schuster, and S. Sutton (London: Routledge, 2000). Many of McLaughlins conclusions, especially in section 4.2, parallel the conclusions reached below.

[8] For a thorough comparison of Descartes' and Newton's theories of inertial motion, see, for example; A. Gabbey, 1980, 230-320.

[9] As R. S. Westfall has pointed out, there is a great deal of ambiguity in Descartes' use of the term "agitation." Yet, Westfall agrees that this term seems to signify "momentum" (or quantity of motion, if its a scalar property) in the articles on the motions of stars in Pr III of the *Principles.* See, R. S. Westfall, *The Concept of Force in Newton's Physics* (London: MacDonald, 1971), 61-62.

[10] This interpretation also accords with the separate role that surface area plays as regards vortex motions. In Pr III, surface area is responsible for the force exerted on an object while immersed within a circling mass of plenum particles, "because the larger [a body's] surface is, the greater the quantity of matter acting against the surface (Pr III 121)."

[11] Descartes' account of "fluid" bodies seems also to confirm these conclusions. In Pr II 58, he remarks that many typical fluid bodies, such as air and water, put up great resistance to the rapid motions of bodies. Thus, in the collision rules, his ideal condition for the absence of plenum effects apparently translates into an appeal for a resistless "perfect fluid." E. J. Aiton, 1972, 39-41, makes a similar observation.

[12] Volume and surface area are two distinct quantities that can be easily confused as equivalent: For example, the volume of a sphere equals the cube of the radius times 4/3 π, while its area equals the square of the radius times 4π.

[13] D. M. Clarke is a notable exception. (1982, 213-221), for he correctly points out both the function of solidity and the elimination of surface area via the idealized conditions. However, Clarke fails to mention the crucial passages in Pr III of the *Principles,* especially Pr III 123 (see above). Thus he does not clearly draw together the three bodily properties of volume, quantity of matter, and surface area as utilized by Descartes.

[14] For example, see: M. Jammer, *Concepts of Mass in Classical and Modern Physics* (Cambridge, Mass.: Harvard University Press, 1961), 60-61; R. J. Blackwell, "Descartes' Concept of Matter", in *The Concept of Matter in Modern Philosophy* (Notre Dame: U. of Notre Dame Press, 1963), 69, fn. 18; J. B. Barbour, *Absolute or Relative Motion?,* vol. 1, *The Discovery of Dynamics* (Cambridge: Cambridge University Press, 1989), 429; P. Damerow et al., *Exploring the Limits of Preclassical Mechanics* (New York: Springer-Verlag, 1992), 76.

[15] J. Carriero, "Comments on E. Slowik's 'Perfect Solidity, Quantity of Motion, and the Problem of Matter in Descartes' Universe' (an earlier version of the present paper)," which were both delivered at the American Philosophical Association Central Division Meeting, Chicago, Spring 1995.

[16] In general, the erosion and wear of a body's surface, which are rather complex phenomena, will result in a loss of kinetic energy. R. M. Brach, *Mechanical Impact Dynamics: Rigid Body Collisions* (New York: John Wiley & Sons), 119-120.

[17] R. Descartes, *The Philosophical Writings of Descartes, vol. 3, The Correspondence,* eds. and trans. J. Cottingham, et al. (Cambridge: Cambridge University Press, 1991), 246-248. This point is raised by P. Damerow, et al., 1992, 102.

[18] Leibniz was one of the first commentators to draw attention to this problem. See, G. W. Leibniz, "Critical Thoughts on the General Part of the Principles of Descartes," in *Leibniz: Philosophical Papers and Letters* (Dordrecht: D. Reidel, 1969), 403-407. Moreover, a requirement for congruent contact surfaces can be found in work of another Cartesian, William Neile (1637-1670): "The whole square surface of the one [cube] meets in the same instant of time with the whole square surface of the other." See, W. Neile, "Hypothesis of Motion," in *The Correspondence of Henry Oldenburg, Vol. 5,* eds. A. R. Hall and M. B. Hall (Madison: University of Wisconsin Press, 1968), 519-524. This is also noted by P. Damerow, et al., 1992, 102.

[19] See, W. L. Scott, *The Conflict Between Atomism and Conservation Theory 1644-1860* (London: MacDonald, 1970), 14, 23.

[20] G. de Cordemoy, *Oeuvres philosophiques.* P. Clair and F. Girbal, eds. (Paris: Presses Universitaires de France, 1968), 103-104; J. Rohault, 1969, 28; B. Spinoza, *Ethics,* Part I, Prop. 15., Demonstration, in *Spinoza Opera,* ed. by G. Gerhardt (Heidelberg: C. Winter, 1925).

[21] G. (Rodis-)Lewis, *L'individualité selon Descartes.* (Paris: J. Vrin, 1950), 45-57; D. Garber, 1992, 175-176; D. Des Chene, *Physiologia: Natural Philosophy in Late Aristotelian and Cartesian Thought.* (Ithaca: Cornell University Press, 1996), 367-377.

[22] S. V. Keeling, *Descartes,* 2nd ed. (Oxford: Oxford U. Press, 1968), 145; R. S. Woolhouse, 1993, 22-23, 55; W. E. Anderson, "Cartesian Motion", in *Motion and Time, Space and Matter: Interrelations in the History and Philosophy of Science,* ed. by P. K. Machamer and R. G. Turnbull. (Columbus: Ohio State University Press, 1976), 202; G. Rodis-Lewis, *L'oeuvre de Descartes.* (Paris: J. Vrin, 1971), 548; D. Marshall, jr., *Prinzipien der Descartes: Exegese.* (Freiburg: Verlag Karl Alber, 1979), 54-57. Listing all the commentators who accept the monist position, or at least find the notion of individual corporeal substance untenable, would take up nearly the entire length of this essay.

[23] It should be noted that Descartes' definition of substance, as "independence from other substances", does not exactly correlate with substance as understood by many of the Scholastics, who often held a much more elaborate notion (e.g., Toletus). On this topic, see, D. Des Chene, 1996, chap. 3.

[24] As G. Brown has argued, nevertheless, it may not be necessary to read the intimate correlation between extension and matter as a straightforward identity claim; "Mathematics, Physics, and Corporeal Substance in Descartes", *Pacific Philosophical Quarterly,* 70 (1989), 281-302. This issue

will be raised briefly below, although the identity of extension and matter, or lack thereof, will not affect the main arguments of this essay.

[25] In his correspondence on this thought experiment, Descartes makes it clear that the removal of the body in the vessel results in the sides becoming "immediately" contiguous, but that there is no motion involved in this process (which would imply the contradictory state of affairs, for Descartes, of the sides moving into the empty space left behind). See, AT IV 109, AT II 482. Some commentators have failed to recognize this crucial distinction, and have accordingly posited motion to the sides of the vessel after the removal of the body: for example, J. Rohault, *ibid.,* vol. 1, 28; and M. Jammer, *Concepts of Space,* 2nd ed. (Cambridge, MA: Harvard U. Press, 1969), 43-44.

[26] Spinoza is not a straight-forward Cartesian, of course, since many of his own ideas differ quite drastically from Descartes' (especially on the metaphysical problem of dualism). Yet, their respective ideas concerning natural philosophy are similar enough to justify the general label.

[27] G. W. Leibniz, "Letter to De Volder, May 13, 1699", *in G. W. Leibniz: Philosophical Papers and Letters,* trans. and ed. by L. E. Loemker. (Dordrecht: D. Reidel, 1970), 519.

[28] M. Gueroult, "The Metaphysics and Physics of Force in Descartes", in *Descartes: Philosophy, Mathematics, and Physics,* S. Gaukroger, ed. (Harvester Press: Sussex, 1980) 199. Gueroult's full presentation of his argument appears in *Descartes' Philosophy Interpreted According to the Order of Reasons, Vol. II,* trans. by R. Ariew (Minneapolis: U. of Minnesota Press), 63-74.

[29] J. Bennett, "Spinoza's Metaphysics", in *The Cambridge Companion to Spinoza,* ed., D. Garrett. (Cambridge: Cambridge U. Press, 1996), 70-71. Interestingly, Bennett's interpretation, which he dubs a "field" theory, is much in the spirit of later "supersubstantivalist" theories of space (i.e., where space is the only predicable substance; and thus objects are merely qualitatively/quantitatively distinct portions of space). Newton toyed with this idea in his early essay, *De Gravitatione et aequipondio fluidorum,* trans. and eds. A. R. Hall and M. B. Hall, in *Unpublished Scientific Papers of Isaac Newton.* (Cambridge: Cambridge U. Press, 1962), 139. We will return to this subject in chapter 6.

[30] T. M. Lennon, *The Battle of the Gods and Giants: The Legacy of Descartes and Gassendi.* (Princeton: Princeton University Press, 1993), 204-205.

[31] AT VII 14. *The Philosophical Writings of Descartes, Vol. II,* trans. and ed. By J. Cottingham, R. Stoothoff, and D. Murdoch (Cambridge: Cambridge U. Press, 1984), 10.

[32] This problem with the modal thesis is raised by P. Hoffman, "The Unity of Descartes' Man", reprinted in *The Rationalists,* ed. by D. Pereboom (Lanham: Rowman & Littlefield, 1999), 64.

[33] Ironically, Descartes' "empty vessel" argument can be seen as an early forerunner to the "hole" problem that plagued Einstein's development of General Relativity, as well as recent formulations of substantivalist spacetime theories. See, J. Earman, *World Enough and Space-Time.* (Cambridge, MA: MIT Press, 1989), chap. 9.

[34] E. Grosholz raises related difficulties, stressing that Descartes' arguments against the vacuum can only work for finite, bounded vacua, and not necessarily against a vacuum that surrounds the material world: "Descartes and the Individuation of Physical Objects", in *Individuation and Identity in Early Modern Philosophy,* ed. by K. F. Barber and J. J. E. Gracia (Albany: State U. of New York Press, 1994), 48-53. Also, Descartes' term "indefinite" essentially functions as "infinite", since it stands for a concept of which no limit can be conceived. Only God, however, is the proper subject of the "positive"-sounding designation, "infinite", which lacks any sense of limit (Pr I 26-27).

[35] E. Grosholz, *Cartesian Method and the Problem of Reduction.* (Oxford: Clarendon Press, 1991), 68.

[36] There were many sixteenth and seventeenth century Natural philosophers who did accept a theory with similar implications, such as F. Suarez, P. Fonseca, the Coimbra Jesuits, B. Amicus, and E. Maignan. These thinkers tried to accommodate both the notion of a finite extended world and an infinite (quasi-) non-extended void: i.e., an "imaginary" space that, while not possessing extension or dimensionality, had the capacity to receive extended bodies. Descartes' theory of space rules out this option, for he interprets vacuum as an extended "nothing", which is an impossible state-of-affairs (since "nothing" cannot be the subject of a property). Thus the actual existence of a plenum is impossible since extension must be a property of some thing (but this should not be taken to undermine the argument centered upon conceivability that will be advanced below). (Pr II 16) His substance-property dichotomy is clearly at work, here. (see, Pr I 56) On "imaginary" space theorists, see, E. Grant, *Much Ado About Nothing: Theories of Space and Vacuum from the Middle Ages to the Scientific Revolution.* (Cambridge: Cambridge U. Press, 1981), chap. 7.

[37] R. Florka has suggested to me that a better interpretation of the consequences of removing all of the space surrounding a body is that it would now fill all of extension: i.e., the body, say, a chair, would now assume the role of the "indefinitely" extended plenum, which is a better analogue to the case of the empty vessel, as examined above. Of course, the problem that would now arise is whether or not the object retains its identity

after it becomes indefinite extension; since, if it is indefinitely extended, and thus has no boundary, in what sense can it still be referred to as a "chair"? Overall, both versions of individual material substance, i.e., Florka's and my own, face the difficulty of coherently explaining the boundary, or surface, of these substances given their lack of contiguous neighbors, thus both approaches to this problem can be seen as complimentary. R. Florka, "Comments on Slowik's 'Descartes, Monism, and Individual Corporeal Substance'", presented at the American Philosophical Association Central Division Meeting, New Orleans, May 8, 1999.

[38] M. Grene, *Descartes*. (Minneapolis: U. of Minnesota Press, 1985), 100-101.

CHAPTER 5

QUANTITY OF MOTION: THE ORIGIN AND FUNCTION OF THE CARTESIAN CONSERVATION PRINCIPLE:

This chapter picks up the story, began in the previous two chapters, concerning the complex and varied role of Descartes' conservation law for the "quantity of motion" (size times speed), the numerical measure of "force" that, he believes, is conserved in all bodily interactions. After presenting the historical development of this concept in Descartes' work, this chapter will explore in more detail various problematical aspects of Descartes' conservation principle noted in chapter 3 and 4; namely, its largely neglected "dual role" as a measure of both durational motion and instantaneous "tendencies towards motion". Overall, an underlying non-local, or "holistic", element of quantity of motion (largely derived from his statics) will be revealed as central to a full understanding of the conservation principle's conceptual development and intended operation. This insight into the function of the Cartesian conservation principle will not only be of use in responding to some of the criticisms of Descartes' physics aimed at its overall consistency, such as Alan Nelson's recent allegations, but will also provide a potentially powerful means of constructing a successful Cartesian spacetime, as will be later demonstrated in Part III (chapter 6).

5.1. Quantity of Motion: The Contemporary Opinion.

While it is commonplace to praise Descartes for having brought conservation principles to the forefront of natural philosophy, his particular choice of a conservation principle of moving bodies has generally received a less enthusiastic welcome. That is, even though the fruitfulness of the conservation "idea" was quickly perceived (and can be said to largely justify, say, C. Truesdell's judgment that "[Descartes' physics] is the beginning of theory in the modern sense"[1]), it is nevertheless also true that Descartes' handling of his conserved "quantity of motion", or size × speed, was soon challenged by a host of natural philosophers, both Cartesian and non-Cartesian.[2] Much of the early criticism centered upon the apparent incompatibility of the Cartesian conservation principle, and its attendant natural laws, with the basic operation or structure of Descartes' matter-filled universe (or, plenum). This line of critical analysis is still pursued by many commentators; such as, A. Nelson, who has charged that the very concepts of Cartesian matter and motion preclude the successful establishment of Descartes' third natural law, and by implication, the conservation principle.[3]

One of these puzzles, to be more precise, is the apparent dichotomy in Descartes' handling of bodily motion. On the one hand, he stipulates that motion is merely the reciprocal translation of a body from the vicinity of its contiguous neighbors (Pr II 25), while, on the other hand, also insisting that God maintains an unchanging measure of motion in the universe (i.e., the conservation principle). But, how do these two descriptions, or hypotheses, interrelate? On the first hypothesis, the complexity of the entire range of plenum activity seems to be reduced to, and explained by, the *local* level of a relational change of "place". In contrast, the second hypothesis explains the local motions and collisions of bodies from the *global* perspective of a quasi-"property" conferred upon the plenum via God's "immutability" (Pr II 36): i.e., QM is a consequence of God's "sustaining act". In short, one approach is local, reductivist, and kinematic; while the other is global and metaphysical (or, non-reductive and non-kinematic). This underlying tension in Descartes' system can be discerned in the work of various commentators on Cartesian motion, although it is often not openly or deliberately acknowledged. In an influential essay by R. J. Blackwell, for instance, it is observed that "everything in the [Cartesian] universe other than extension is due to local motion."[4] In this same work, nevertheless, he concludes that "for Descartes the word 'conservation' [in conservation principle] refers more to something in God (i.e., immutability) than to anything properly inherent in the structure of the material universe (223)." Once again, how can motion be an entirely local affair, but also such that "if there were no God, the total quantity of motion may well not be conserved in the Cartesian universe (234)?"

Blackwell's analysis is typical of the work on Cartesian motion, and nicely demonstrates the subtle ontological and epistemological issues invariably raised by such investigations, both with respect to corporeal substance and God's sustaining act. This chapter will attempt to come to grips with at least one aspect of this complex metaphysical problem: the local versus the global, or reductivist versus non-reductivist, nature of Cartesian motion. Despite being long favored by commentators, as well as seemingly endorsed by numerous passages in Descartes' corpus, it will be argued that the local, reductivist reading just cannot make sense of the entire Cartesian conception of motion. A global, non-reductive, and somewhat "holistic" element pervades Descartes' thinking and approach—and the conservation of quantity of motion is the key to unlocking this unacknowledged aspect of his theory. In particular, Descartes' early work on statics will be of central importance for our investigation, since the quasi-holistic character of a system of bodies held in an equilibrium of forces, such as weights on a pulley or balance, retained a pervasive influence on his later approach to the problem of motion and force in the plenum. The blending of this statics outlook with his kinematical set of natural laws inevitably generated numerous difficulties and inconsistencies for Descartes' system, as will be in evidence throughout our investigation. Nevertheless, acknowledging the dichotomous nature of Descartes' conservation principle for the quantity of motion will be shown to be oddly useful in the debates surrounding some of the alleged difficulties with Cartesian natural philosophy—and, more importantly, it will have a corresponding beneficial effect on our primary goal of developing a consistent Cartesian spacetime.

(Henceforth, 'QM' will designate quantity of motion, and not the general notion of a conservation principle, which can take any number of forms different from "size times speed".)

This chapter will bring together many seemingly disparate issues. In section 2, we will examine the origins and structure of Descartes' universally conserved motive quantity. This analysis will help to clarify the ontological status and unifying importance of QM in his universe, as well as its dual instantaneous/durational and dual static/dynamic character. In section 3, the conservation law will be put to work in order to resolve a series of alleged difficulties with Cartesian matter in its plenum setting. The aforementioned interrelated themes of reduction and holism will be a principal component of this investigation, and will be shown to play an important part in the relationship between Descartes' problematic thesis of local, relative motion and his conserved force.

5.2. Quantity of Motion in Cartesian Natural Philosophy

As D. Garber has carefully documented, Descartes' conservation principle only reached its canonical formulation, in his *Principles of Philosophy* (1644), after a long process of development and gestation. In his earlier, suppressed *The World* (1633), Descartes claims that God "always conserves [in all of matter] the same amount of [motion] (AT XI 43)." But it is not clear, at least in this work, that QM (size × speed) is the "motion" conserved.[5] More than likely, the source of Descartes' interest in conserved properties of motion was his friend and mentor, I. Beeckman, who formulated an early version of the momentum law, or size × velocity.[6] Regardless of the source of his inspiration, however, Descartes' references to a conserved QM become more explicit by the late 1630's (e.g., see AT II 543), culminating in an unambiguous presentation in the *Principles:*

> [Motion] has a certain and determinate quantity, which we can understand easily to be able to remain always the same in the whole universe, even though it may change in its individual parts. That is why we might think that when one part of matter moves twice as fast as another which is twice as large, there is the same amount of motion in the smaller as in the larger (Pr II 36).

Of course, in this treatise, the conservation principle forms the foundation of the three laws of nature, which describe the basic motions of material bodies in the Cartesian plenum. As noted in chapter 3, QM is particularly conspicuous in the third natural law, and the accompanying seven collision rules which constitute its specific instances, since this law dictates that impacting bodies either retain all their motion or transfer as much motion as they lose (and thus conserve QM; Pr II 40). Descartes justifies the conservation principle through an appeal to God's "immutability," as he dubs it (Pr II 37). He states: "It seems obvious to me that it is none other than God himself, who created in the beginning both motion and rest in the whole of matter, and now preserves through his normal concourse alone the same amount of motion and rest as he placed in it at that time (Pr II 36)."

Yet, to a host of Cartesian scholars reared on the "a priori" epistemology of the *Meditations,* the form of Descartes' conservation law (for QM) must appear a rather unmotivated and problematic choice. Whereas other Cartesian hypotheses could conceivably have a rational foundation based on God's immutability (albeit improbably; e.g., extension as the sole attribute of matter, Pr II 11, and even the existence of a conserved property of motion), it is not at all obvious, in the a priori sense, why the totality of plenum motion must equal the size × speed of all plenum bodies. Why this particular quantity? Why not "size + speed", or "size × speed^2"? In short, Descartes' conservation principle incorporates two key assumptions which demand justification: first, the choice of simples or primitives ("size", "speed"), and second, the mathematical relationship that ties the primitives together (product function, " × "). As for the domain of primitives, Descartes reckons size and speed to be "modes" of extended substance (Pr II 27); which, in the *Principles* at least, are defined as the particular ways in which a corporeal body manifests its extension. Size and speed thus appear to fit into the Cartesian ontological scheme, since they are a natural correlate of his a priori identification of extension as matter's essential attribute. There are problems with this selection, however. Not only is Descartes' concept of "size" notoriously difficult to interpret, but other presumably basic properties of Cartesian matter and motion are excluded without explanation, such as the direction of motion (or, approximately, "determination").[7] A similar puzzle surrounds the choice of the product relationship, ×, to interrelate his simples. On the whole, it does not appear that any a priori justification can be given, based on God's immutability, for privileging the product function over the range of all possible quantitative relationships (e.g., addition, subtraction, division, etc.). This dilemma is nicely summarized by S. Nadler: "It is plausible to claim that [God's] immutability implies *some* conservation law. But that it implies the conservation of the total quantity of motion in the universe is questionable."[8]

If Descartes did not procure QM based on any sort of a priori reflection, on what grounds did he select this particular quantity (and from where did it originate)? The most plausible explanation is to assume that Descartes simply borrowed this quantity, which was well-known to his predecessors in both statics and impetus theory, and generalized it to cover all material interactions in his plenum. Descartes probably knew the many Scholastic theories grouped under the general title "impetus" from his early Jesuit training in natural philosophy. Recalling our discussion from chapter 3, impetus theories were developed as a sort of a forerunner to the modern concept of inertial motion, and Descartes' own laws of motion. A body's impetus was conceived as a force or quasi-property causally responsible for its "violent" motion away from its naturally resting terrestrial state (i.e., naturally at rest when confined to the surface of the earth). This force was often measured by the speed or velocity of a body and its quantity of matter, as in the case of John Buridan.[9] In statics, further quantities bearing a close resemblance to QM were the various dynamical renditions of the "law of the lever," later dubbed in their mature form, the principle of "virtual work." Statics is the branch of mechanics that examines the forces of bodies in equilibrium, such as weights suspended in a pulley or balance. Descartes was well-acquainted with the theoretical aspects of these types of simple machines, as is evident in several letters from the late 1630s. For example:

> To raise a 100-pound weight to a height of one foot twice is the same as raising a 200-pound weight to a height of one foot, and also the same as raising a 100-pound weight to the height of two feet. From this it clearly follows that . . . the force required to support it at a certain position and prevent it from falling, is to be measured at the beginning of the motion which the power supporting it must provide if it is going to raise it or if it is going to accompany its falls (AT II 229—see Figure 17).

The measured quantity in Descartes' rendition of the law of the lever incorporates only very small displacements (along with weight), thus rendering it roughly analogous to the later virtual work principle: "Note that I say 'begin to fall', and not simply 'fall', since it is only the beginning of the fall [of the weights] that we need to consider (AT II 233)."[10] Following A. Gabbey, we will label Descartes' treatment of the lever law, the "General Statical Principle," or GSP.[11] The similarities between QM and the GSP are quite evident, as are their differences. Despite the obvious parallel use of the product function, the GSP's elements are weight and infinitesimal displacement (where "infinitesimal" simply denotes "minute" or "small", and not its later mathematical usage). QM, on the other hand, employs size and non-instantaneous speed (Pr II 39; see below).

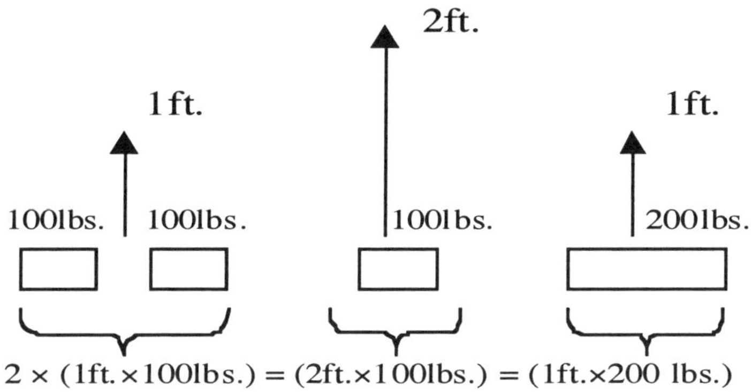

Figure 17. Three different instances of raised bodies with identical values for the quantity of motion.

Nevertheless, the differences between QM and the GSP are not as great as they might at first seem. In one sense, of course, QM is non-instantaneous; i.e., it includes the Cartesian concept of speed, which is only manifest over a non-instantaneous temporal period (as best exemplified in the collision rules, Pr II 46-52). Yet, QM is also closely tied to the instantaneous property[12] which we can loosely entitle, "tendency" (*tendere,* Pr II 39), although the *Principles* employs several designations interchangeably in depicting this notion (e.g., "striving," *conatus,* Pr III 56; "first preparation for motion," *prima praeparatio ad motum,* Pr III 63). To summarize the conclusions from chapters 3 and 4 above: tendencies are involved or implicated in the (non-instantaneous) motions of bodies, and thus are carefully coupled to the states of affairs presiding at a single instant. He states: "Of course, no motion is accomplished in an instant; but it is obvious that every moving body, at any moment in the course of its motion, is determined (*determinatum*) to persevere in its motion in some direction along a straight line, . . . (Pr II 39)." Given the context of this passage, Descartes seems to imply that a tendency has a determination (which roughly corresponds to the direction of motion, as above), for he also states in Article 39 that "each part of matter, . . . tends to persevere in its motion along only straight lines. . . ."[13] Furthermore, in discussing the motions of stars, he applies the name "agitation" (*agitation*) to signify the "force," or measure of the force, of these tendencies: "Once moved, gold, lead, or other metals retain more agitation, or force (*vis*) to persevere in their motion, than pieces of wood or rocks do of the same size and shape (Pr III 122)." The descriptions of the agitation force in this part of the *Principles* exactly match his earlier definition (in Pr II) of a moving body's QM, which is a measurement of its "force to persevere (*perseverare*) in its motion, i.e., in motion at the same speed in the same direction (Pr II 43)." QM is hence a gauge of agitation force, while the agitation force, in turn, is a measurement of the instantaneous tendencies towards motion exhibited by Cartesian bodies.[14] To summarize, QM would appear to serve two primary functions: first, it is a measure of non-instantaneous size × speed; and second, it is a measure of instantaneous tendencies towards motion (see chapter 4 for more details). With regards to instantaneous quantities in general, there is thus a close parallel between QM and the GSP.

In fact, bearing in mind that Cartesian motion does not take place at the level of instants, it becomes difficult to draw a sharp distinction between instantaneous tendencies towards motion, as measured by QM, and infinitesimal displacement, as incorporated in the GSP. If instantaneous tendencies towards motion are not "real" motions, in Descartes' sense, should we consider them to be more like an infinitesimal displacement (change of place) than a motion? We are venturing into uncharted waters, here, since there is ostensibly nothing in the Cartesian corpus to resolve such debates (although suggestions will be offered below). Descartes never fully addresses the many issues and problems that pertain to infinitesimals, and his *Géométrie* eschews all reference to them.[15] As mentioned, it is an open debate as to whether he conceived instants of time as absolutely instantaneous or including some temporal duration. Therefore, the relationship between tendencies and minute displacements remains mysterious, if yet somehow closely connected given both their infinitesimal nature and the close connections between motion, displacement, and time.

As for gravity or heaviness (*gravitas*), however, it can be stated with complete assurance that it is closely related to the Cartesian notion of size. The motion of bodies towards the center of the earth is a direct function, or consequence, of bodily volume, surface area, and quantity of second and/or third element matter; i.e., size. It is only in Prs III and IV of the *Principles* (and not in the presentation of the natural laws in Pr II), that the complex interplay of these quantities is fully disclosed, along with their relationship to the natural (inertial) motions of bodies.[16] All in all, gravity is simply the process whereby slow moving terrestrial bodies are "pushed down", in a contact-mechanical fashion, by the faster moving plenum particles which rise to take their place. The greater "force" of motion of the plenum particles—i.e., their tendency "to persevere in" straight line uniform motion, as measured by QM—is the sole cause of the descent of terrestrial bodies, as well as the increasing velocity, or acceleration, of their free-fall over time (due to the incremental build-up of each separate "push"). Shape, volume, surface area, etc., play a role in determining the force of these bodily motion, consequently, since Cartesian matter is ultimately pure extension. In short, weight is as closely linked to size and speed, as instantaneous tendencies and QM are related to infinitesimal displacements and the GSP.

The moral of the story, thus far, is that Descartes largely based his conservation principle of QM on his earlier work in statics and the GSP, with an additional contribution possibly stemming from his knowledge of Scholastic impetus theories (especially those which employ QM-like quantities). The purported statics background to Descartes' physics appears to have been called into question, however, by S. Gaukroger's investigation of the early hydrostatics tract (of 1618, AT X 67-74).[17] In short, Gaukroger regards Cartesian physics as more "hydrostatic" in nature, as opposed to kinematic (or some other branch of mechanics), given the presence of the many fluid motion metaphors in *The World* and *Principles*. While it is true that Cartesian physics is often presented within a hydrostatics context, especially in the early works, Descartes' overall treatment of hydrostatics is decidedly statical. The evidence for this conclusion lies in Descartes' particular handling of the traditional hydrostatics problems, such as fluid pressure depending on height and not area, from the perspective of the instantaneous tendencies of single fluid particles to descend (see, Figure 18). As W. R. Shea likewise observes, Descartes' hydrostatics approach thus appears to be based on his previous knowledge of statics, although the source of Descartes' knowledge in this area is uncertain.[18] (If this point is conceded, however, then Gaukroger is probably correct in arguing that Descartes' understanding of hydrostatics largely shaped his later plenum model, since this would explain its latent statical characteristics.)

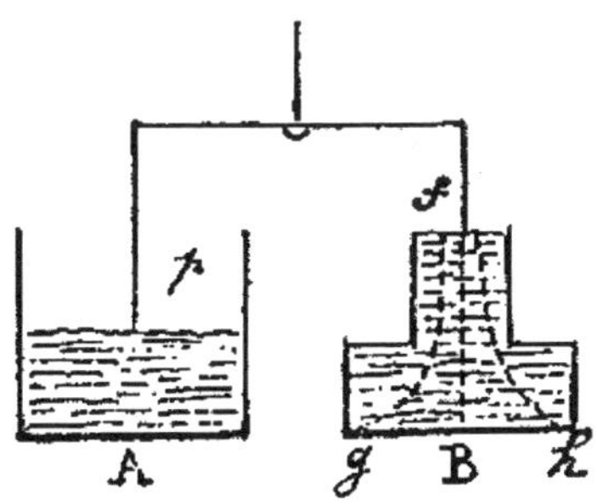

Figure 18. In Descartes' early notebook (circa 1618, AT II 228), the fluid pressure on the bottom of container B is analyzed in terms of the instantaneous tendencies of the "atoms" of water to descend along various paths, labeled *g*, B, and *h*.

Our statics-based assessment of the genesis of Descartes' ideas is not particularly unique or controversial, it should be noted, for similar conclusions have been advanced by previous commentators (see footnote 15). Westfall, for one, has commented that:

> size times velocity [or speed], derived from the law of the lever, bedeviled mechanical discussions throughout the century as scientists attempted to apply to real motions a quantity valid for virtual velocities [weight times infinitesimal motion]. Descartes also employed the concept as the force of a body's motion, and through that concept the statics of simple machines continued to dominate dynamics (Westfall 1971, 75).

Needless to say, such estimations are merely conjectural due to the lack of any overt reference in Descartes' work as to the source of his QM. But the "weight" of the evidence based on Descartes' scholastic training and his earlier written work strongly favors a GSP/QM correspondence, as Westfall suggests.

Granting this point should not lead one to conclude, however, that the GSP is somehow foundationally prior to QM as regards Descartes' overall ontological and epistemological scheme. He may have decided to base QM on his experience with the GSP, but the three laws of nature, including the conservation principle, remain the "a priori" cornerstone of his entire natural philosophy. The mechanics of simple machines and the GSP are derivative principles on this plan, as is revealed with notable clarity in his September 12, 1638 letter to Mersenne:

> Many people often confuse consideration of space [distance] with that of time or speed. . . . For it is not the difference in speed that makes it such that one of these weights has to be twice the other [with respect to a balance where the arm of the smaller weight is twice the length of the heavier weight], but the difference of space [i.e., distance through which the weights rise or descend]. . . . If I had wished to link considerations of speed to that of space, I would have had to attribute three dimensions to the force, instead of the two I had attributed, in order to exclude [speed]. . . . For it is not possible to say anything good and sound with respect to speed, without first having explained what weight [gravity] really is, and simultaneously the whole system of the world (AT II 353-355).

In other words, whereas the force measured by the GSP mandates two "dimensions," weight and distance, the force of bodily motion under the influence of gravity (in raising bodies or in free-fall) requires the added dimension of speed. To adequately account for gravity, he continues, it is necessary to explain first his entire vortex system (Pr III and IV of the *Principles*); which, respectively, is grounded upon the Cartesian natural laws (in Pr II). The bodily forces involved in gravity, and the corresponding dimensions of weight, distance, and speed, thus depend (in a roundabout way) on QM. In essence, this conceptual scheme places QM at the ground level of Cartesian physics, since the GSP relies on the explanation of weight that only QM, via the vortex theory, can provide.[19]

This assessment of the relative positioning of Descartes' physical principles has been duly noted by A. Gabbey, whose work in this area of Cartesian science has set new standards of scholarship. Gabbey draws a further inference, however, which we will need to address before proceeding to the next section. He states:

> The GSP, Descartes' version of the work principle, is not a *lex* or *principium physicae* in the sense of being irreducible to other than metaphysical considerations. It is a lawlike empirical rule that explains the operation of devices under the inviolate influence of gravity. For Descartes the irreducible *leges physicae* are centrally the three laws of nature set out in Part II of the *principles,* and they are necessarily true of any divinely created possible world containing *res extensae* in motion. On the other hand, the GSP is contingently the case for the actual world God has chosen to create (1993, 320).

Gabbey seems to believe that, unlike the natural laws, the GSP is not one of Descartes' "eternal truths" (or, at the least, one of those truths or features of the world that must remain the same in all the worlds God could create).[20] This estimation of the ontological and/or epistemological status of the GSP is not warranted by the evidence, I would argue. Although it is clearly the case that the natural laws possess some form of necessity, Descartes' paltry references regarding the underlying principles of simple machines simply do not rule out the possibility that the GSP is likewise an invariant truth of all divinely created possible worlds. If Descartes would have finished the project started in the *Principles,* and included a comprehensive treatment of mechanics (as confessed in his preface to the French edition, AT IXB 14), this may have decided the matter one way or the other. As it stands, however, there seems as much reason to judge the GSP "a necessary feature of all God's possible worlds" as there is to rule it out.

In fact, given its close kinship to QM, and his analysis of the dimensions of force, it might be more plausible to rank the GSP as a feature of Descartes' world equal in necessity to QM and the natural laws. It would seem odd, for instance, that in moving from the three dimensions of the bodily force of gravity—a system fixed by the necessity of the natural laws—we arrived at a two-dimensional statics level whose corresponding force is "contingently the case for the actual world God has chosen to create". Put briefly, Descartes' statics seems invariably bound to his analysis of gravity; with the latter being, presumably, a necessary consequence of his natural laws. Moreover, even if a different vortex system were compatible with the same natural laws (which might precipitate a different account of gravity and its three dimensions), it is still hard to envision how Cartesian statics could vary independently of Cartesian gravity: as a sort of two-dimensional restriction, or special case, of three-dimensional gravity, there is good reason to believe that what is true at the level of three will also be true at the level of two.[21]

5.3. Quantity of Motion in the Cartesian Plenum.

In the previous section, two distinct aspects of Descartes' QM were discerned: the measure of a body's non-instantaneous, durational, size × speed; and the measure of a body's instantaneous, non-durational, tendency towards straight-line uniform motion. In the present section, we will explore in more depth how QM's dual nature can be exploited to counter a range of perceived difficulties with the Cartesian conservation principle (and QM itself). In particular, section 3.1 will provide a lengthy examination of A. Nelson's critique of the conservation principle; while sections 3.2 and 3.3 will explore the difficulty of upholding the principle given Descartes' theory of relational motion. The main burden of our attempts to resolve some of QM's problems will be carried by, for lack of a better word, a "holistic" component of QM. Yet, as will become evident, the advantages that accrue to a holistic interpretation of QM (in meeting these challenges) cannot ultimately overcome all of the difficulties generated by Descartes' dichotomous local-kinematic/global-dynamic conception of his conservation principle.

5.3.1. Micro-Chaos.

Idealization in Cartesian physics, or its failure, forms the backdrop of Nelson's foray into the vagaries of Descartes' collision theory. Among other things, Nelson contends that the stipulation in the *Principles* for perfectly hard bodies, and for motions unaffected by plenum crowding (Pr II 45), are not sufficient in themselves to permit the successful application of Descartes' natural laws. In order to demonstrate how the holistic properties of QM can be of service in countering Nelson's allegations, it will be necessary to provide first a careful presentation of the many concepts involved. For our purposes, the third natural law, and its accompanying seven collision rules, will be of primary concern, since this law prescribes the conservation of QM for each particular bodily collision (Pr II 40). Nelson argues, with some justification, that it is probably only at the smallest level of Cartesian material substance, the first and second elements, that actual observation of plenum interactions would conform to the predictions of the natural laws. (See, Pr III 52, for the tripartite classification of Descartes' material elements.) At this minute scale, Cartesian elements are not only "harder" than the bodies they comprise (where "hardness" is defined as the lack of relative motion among the constituent parts, 378), but the numerous channels and pores in large bodies probably facilitate their motion more readily, thus guaranteeing a closer approximation to the isolation criterion.

Unfortunately, Descartes' rejection of atomism (Pr II 20), and his belief that first element particles can be "indefinitely small" (Pr III 52), would seem to preclude the realization of perfectly hard material substances, or, to use Nelson's term, "impenetranium". This is only the beginning of the difficulties, however. Any attempt to apply the natural laws to, say, a closed loop of circling bodies bounded by a channel of hard matter (a favorable idealization in its own right), will be inevitably undermined by the complex motions of the first element particles, a process entitled, "micro-chaos".

> Real channels are to some degree soft, and generally in some kind of motion [i.e., they are not impenetranium]. This means that many of the tiny particles of the first element that are being scraped or shorn off the bodies in the circuit [i.e., closed ring of circling matter] will be continually seeping out of the channel forming tiny new circuits of their own. Now none of these tiny circuits will be encased in perfectly uniform channels of impenetranium, so further even smaller particles will be generated some of which will seep out of their channels and so on *ad indefinitum*. There never is a smallest level at which angels could use the laws exactly to calculate how bodies will collide (Nelson 1995, 382)

Put differently, the indefinite divisibility of plenum matter, which Descartes sanctions as a real process (in order to fill-in the vacuities between larger elements, Pr II 34), undermines the application of natural laws *in principle*. No matter what scale or level of Cartesian matter one considers, the natural laws cannot be successfully applied because the material debris or scrapings that naturally result from the bodily motions at level 1—and which form an even smaller layer of particles at level 2—will escape the application of the laws at level 1 (and so on for level 2, and level 3, etc.). "Micro-chaos quite clearly entails that exact calculations of what is happening to any object at the micro level is utterly impossible", and hence "Descartes' laws are every bit as inapplicable at the micro level, even in principle, as they are at the macro-level (382-383)."

Before venturing to discuss how this critique affects the conservation of QM, a few closely related issues need to be discussed. First of all, it is assumed throughout Nelson's article that Descartes' concept of "perfect hardness" (where the terms *solidus* and *durus* are used interchangeably in the *Principles*) only pertains to the degree of relative motion among a body's constituent particles. Based on this interpretation, and Descartes' admission that "all the variety in matter . . . depends on motion (Pr II 23)," he infers that all material bodies must have some internal motion (if not, then gold would be identical to wood, for example). Yet, as argued in chapter 4, there is good reason to view Descartes' perfect hardness idealization more as a means of obtaining a simplified measure of a body's quantity of second/third element matter to total volume than as a stipulation dealing exclusively with the relative motions of particles (or worse, an elastic/inelastic impact property). The evidence of the entire *Principles,* and not just Pr II, strongly favors this "quantity of matter to volume" reading of perfect hardness, especially if the theory of stellar motion detailed in Pr III 120-125 is taken into account.[22] Therefore, Nelson's contention that "given Hardness, . . . , it is conceptually impossible that [Descartes'] laws apply to any bodies not consisting of impenetranium (379)" would seem to miss the point of the perfect hardness idealization. This criterion is aimed at getting a manageable grasp on that most unmanageable of Cartesian notions, "size", as is further evidenced by his stated rationale for the idealizations: i.e., "Our calculations [of QM and the natural laws] would be more *easy* if . . . (Pr II 45, emphasis added)." Descartes does not argue that perfectly hard macroscopic bodies *do* exist; in fact, he seems to suggest that there are no such macro-scopic bodies (Pr IV 132). Rather, immediately *after* defining QM, and *prior* to laying out the collision rules, he argues that the calculation of QM would be made easier *if* the bodies were perfectly hard. The hardness and isolation (from plenum interference) criteria are idealizations in the following sense: as macro level bodies approach perfect hardness, the calculations of QM will become simpler and more accurate; and the less their motions are impeded by plenum crowding, the closer their impact behavior approximates the predictions of the seven collision rules. This reading of the hardness and isolation idealizations would seem to be born out by his admission, just after submitting the collision rules, that "it is very difficult to perform the calculations [due to the interference of other bodies]. . . (Pr II 53)." (Once again, see chapter 4 for a more detailed exploration of these issues.)

One might insist, however, that such idealizations simply cannot be met in Descartes' plenum, and hence the natural laws remain inapplicable to any real, non-idealized situation. This claim may be partially true, but it is hard to see how this problem is unique to Cartesian physics. This last response leads us naturally to our second point concerning Nelson's critique: In short, how do Descartes' idealizations differ in kind from, say, the Newtonian utilization of "point particles" (which Nelson reckons to be a successful instance of idealization in physics)? He states, "for ordinary bodies, we can calculate a pointlike center of mass that enables us to treat ordinary bodies as if they were themselves point masses (380)." The correctness of this Newtonian idealization is grounded on the following thesis: "Since Newton, it has been characteristic of good idealizations that they permit calculations that hold approximately for real systems, and that the calculations become more exact the more closely the real system resembles the idealization (387)." Yet, if our earlier analysis of Cartesian idealization is correct, then this is exactly what Descartes is attempting to do. Not only is a perfectly hard and isolated Cartesian body an effort to approximate a real system (such that "the calculations become more exact the more closely the real system resembles the idealization"), but it seems to be an idealization exactly akin to the equally-unrealizable Newtonian point mass.

Some of the confusion may stem from the fact that Descartes' isolation criterion seems so alien in comparison to the historical examples often exhibited as "correct" idealizations in Newtonian mechanics; namely, the use of point masses to model complex gravitational systems, such as the solar system. Of course, the development of theoretical mechanics, and the success of Newton's gravitational theory, did largely depend on the point mass approach to the interactions of planetary bodies. But, it is also true that the equally-profitable study of the mechanics of continuously extended materials, i.e., continuum mechanics, employed a very different set of idealizations to a more humble (but no less difficult) range of material phenomena, such as elastic solids and fluid flow. The mathematical models of these phenomena often invoke a criterion of "ideal materials" in an endeavor to get a workable grasp on the extremely complex interplay of material forces and factors. For instance, the interaction of mechanical and thermal processes are often minimized, or simply neglected, in order to focus attention exclusively on the statical and mechanical features of the system. Such idealization, moreover, are not point-like, but extend over an entire body or material. In an analogous sense, Descartes' appeal to perfect hardness and isolation can be seen as a crude forerunner to the types of idealization that concern continuum mechanical theorists. As in the modern theory, Descartes is striving to minimize the number of complicating factors involved in the behavior of extended material bodies. Neglecting thermal processes is thus the analogue of neglecting plenum interference: both are impossible per se, but they can approximate the behavior of bodies reasonably well, and thereby provide useful information on their respective systems.

The analogy between Descartes' plenum physics and continuum mechanics should not be taken to extremes, of course, since they differ on matters of more than just historical origin. Yet, there is a vague Cartesian-esque quality to many of the concepts in the modern theory: for example, an elementary textbook comments that "observed macroscopic behavior [of bodies] is usually explained by disregarding molecular [or particle] considerations and, instead, by assuming the material to be continuously distributed throughout its volume and to completely fill the space it occupies."[23] Correlating spatial volume with indefinitely-divisible matter is quite Cartesian, needless to say, but one has to stretch the interpretation of this passage to encompass Descartes decidedly corpuscularian thesis of the three material elements. Even if we concede this last point, there is a sense in which Descartes does disregard the individual, local behavior of bodies (and elements) in favor of a "holistic" treatment of plenum phenomena, much like the continuum theorist's continuously distributed matter. We have already mentioned, in passing, one instance of this type of thinking: i.e., the quasi-miraculous manner in which the first element particles fill in the gaps left in the wake of larger, moving elements. It is hard to see how this global feature of plenum activity could be derived exclusively from the local behavior of individual first element particles, especially when it is recalled that their behavior is governed by the natural laws alone (i.e., rectilinear uniform motion that conserves QM). This problem would seem to have caught Nelson's attention as well, for he offers it as an alternative interpretation of his micro-chaos argument (382-383).

How can Cartesian bodies harmonize their collective motions so effectively when their individual motions are explained *only* by the three Cartesian natural laws? In the absence of any alternatives, the likely response to this question is to invoke Descartes' conservation principle of QM. In other words, the synchronized motions of the first element particles are merely the result, or consequence, of the conservation of QM at each moment. This non-local, or holistic, property of the conservation principle will figure prominently throughout the remainder of our investigation; yet, now that we have prepared the context, we may proceed to demonstrate its potential with regard to Nelson's (first) critique of plenum motion.

All told, the success of Nelson's argument relies on the indefinite divisibility of Cartesian matter: since we can never arrive at a smallest level of particles, we can never guarantee that both the hardness and isolation idealizations are being met, and thus that the natural laws are being upheld. Nevertheless, QM is not only conserved in the plenum as a whole (Pr II 36), but also in each individual collision (e.g., the third natural law: "if a body collides with another which is stronger than itself, it does not lose any of its motion, but if it collides with a weaker body, the quantity of motion it loses is equal to the amount it imparts to the other body," Pr II 40). Accordingly,

QM should be conserved for any given portion of the plenum, as long as the isolated portion is bordered by bodies, or a body, larger than the bodies contained within that portion. This condition might seem odd, but it follows naturally from the third law, and the fourth collision rule (Pr II 40, 49); which holds that a smaller, moving body rebounds without losing any speed after impact with a larger, stationary body. (It will be assumed throughout that the larger body is at rest.) With a larger body lining the isolated section of the plenum, all the collisions against that border by the smaller bodies located within the section will rebound without losing QM. Therefore, QM will be conserved by the totality of bodily interactions within that plenum section. This is the crucial step, for it guarantees (via the conservation principle) that there exists a fixed or determinate value of QM within the plenum portion (but only as long as the bordering body remains in place). Returning to Nelson's argument, although it is true that there is an ever shrinking level of Cartesian particles to examine, the conservation principle ensures that the values obtained from the measurement of QM will *approach a fixed value* with each descending level of particles we measure. That is, if one adds the measure of QM of each decreasing level of particles to the total QM of all the previously measured levels, then one will inexorably draw closer to a limiting numerical value of QM for the entire material contents of the plenum portion. Mathematically, this could be represented as an infinite series that approaches a limiting sum: e.g., $1 + 1/2 + 1/4 + \ldots + (1/2)^n + \ldots = 2$. As in the mathematical case, one never reaches a last member of the series (or a last layer of Cartesian matter), but the series (and plenum portion) have a limiting sum, nevertheless. This interpretation might draw textual support from the fact that Descartes believed the plenum (=space) to be "indefinitely" extended (Pr II 21), despite having a fixed, and presumably finite, value of QM. Descartes' theory is not unique in this appeal to infinite processes, furthermore. In short, the indefinite divisibility of Cartesian matter is no more a problem for Descartes than are the infinite, and equally unintuitive, consequences of many other scientific theories, regardless of their relative success or acceptance; e.g., Newton's theory of gravitation, where a single body is subject to the gravitational attraction of a possibly limitless number of bodies distributed through his infinitely extended space.

5.3.2. Relational Motion and QM.

Consequently, if we bear in mind the global, or holistic, capacity of the conservation principle to conserve QM over an entire region (and not just locally), then we must reject the claim that the Cartesian natural laws are, in principal, inapplicable. Nelson's inquiry is nevertheless valuable in bringing to light the monumental difficulties that beset the determination of QM given the quirks in the Cartesian scheme. Foremost among these difficulties is Descartes' handling of relational motion, a subject to which we can now finally return (although Part III will take up this topic at greater detail). As described in Part I above, Descartes defines "external place" (Pr II 13) as the boundary between an object's surface and the surface of its contiguous neighbors, which entails relationalism when conjoined with his two further theses that: motion *"is the transfer of one piece of matter or of one body, from the vicinity of those bodies immediately contiguous to it [roughly, external place] and considered at rest, into the vicinity of others* (Pr II 25);" and thus "we cannot conceive of the body AB being transported from the vicinity of the body CD without also understanding that the body CD is transported from the vicinity of the body AB (Pr II 29)." As also described in Part I, Descartes' natural laws apparently contradict this relational view by attributing individual, determinate states of motion to individual bodies (as in the second law of motion).

Before turning to the obstacles of conjoining the conservation principle and a relationalist construal of motion, one of the chief advantages of this marriage should be duly acknowledged. It has occasionally been charged that, unless egregious ad hoc reasoning is introduced, Descartes' theory cannot prevent the possible formation of a single, synchronous plenum motion.[24] On this scenario, although QM is conserved, all the plenum occupants have entered into a synchronized motion of identical speed and direction (presumably circular). The fact that Descartes' natural laws cannot rule out this disturbing possible state-of-affairs, which would effectively eliminate all diversity in the universe, is hence regarded as symptomatic of the untenable character of his basic premises. Oddly enough, while the non-durational "tendency" component of QM may be susceptible to this objection, the coupling of the durational measure of QM and relational motion would seem to be immune. Since (non-zero) QM determined over time requires motion, and motion is a change of relative neighborhood, the conservation principle mandates a total value of QM that accrues from variant individual bodily motions, and not, moreover, from a motion of the plenum en masse. In other words, a mass plenum motion is no motion at all, at least for a relationalist of Descartes' stripe. When all bodies move identically in this manner, there is no change of contiguous neighborhood, thereby entailing a total QM value of zero—which is in direct violation of the conservation principle. This line of reasoning is reminiscent of Leibniz' argument against Newton's absolute space, and may constitute one of the few advantages of Descartes' adopted relationalism.[25]

Nevertheless, the overall relationship between Descartes' relationalism and his conservation principle is quite puzzling, prompting Westfall to comment (among others): "For any two bodies, the quantity of motion varies with the frame of reference; if motion is in fact relative, quantity of motion cannot be an absolute value." (1971, 151) Westfall is surely justified in questioning the relational compatibility of the Cartesian natural laws.[26] The problem with this line of inquiry, however, is that it assumes, perhaps anachronistically, that Descartes' QM must correspond to the modern notion of an "invariant" quantity; i.e., as a sort of kinematic property reducible to the individual, durational motions of bodies before and after impact.[27] Put differently, the universal value of QM can be *reduced* to the sum of each individual body's size × speed (which is why the arbitrariness of relational determinations of motion undermines QM). While this type of thinking may be more typical of later theories, such as Newton's momentum law (size × velocity), it does not seem to capture fully the multi-faceted Cartesian use of this concept. For Descartes, as disclosed in section 2, QM has two distinct, if interrelated, personalities: an instantaneous determination of a body's perseverance in straight line uniform motion, and a durational measure of a body's size × speed. The latter notion may have stemmed from his knowledge of Scholastic impetus theories, but the former concept appears wedded in his statical training. When the force of a body to descend is balanced against that of another, as in simple machines, the resulting quantities or "invariants" of the system take on an infinitesimal or instantaneous form (as in the case of the virtual work principle). What is important to note about these quantities is that they require a system of bodies whose interactions are interconnected or harmonized in a kind of "holistic" fashion. This is not the type of model that is readily derived from experience of the kinematical impact of individual moving bodies, as employed by Newton's, or even Descartes', collision rules. Rather, the arrangement of bodies in a balance will only allow constrained motions of the entire assemblage as a whole: i.e., to analyze the motion of one body, you have to indirectly take into account the movements of all bodies—such that, if one descends, the other rises (at least hypothetically via the law of the lever, and for minute distances or speeds). Consequently, while its exact relationship to relationalism remains inherently problematic, Descartes' instantaneous version of QM implies a system whose overall conserved force cannot be simply reduced to the isolated, individual motions of bodies.

In an ironic twist, the instantaneous tendencies of Cartesian bodies and God's sustaining act are thus similar in their holistic implications. Just as God's role in conserving QM cannot be simply reduced to bodily motion (since God's actions are not mere material phenomena), Descartes' instantaneous version of QM depends on a view of plenum interactions that is equally irreducible to individual bodily motion (since one has to examine the entire interconnected system). Descartes' next crucial step, some might say fatal, is to extend this holistic interpretation of instantaneous tendencies to cover the durational, size × speed, motions of bodies, as well. Much of the analysis of motion in the *Principles* exhibits this statics-based approach. For example, he summarizes at the end of Pr III that "for as all bodies in the universe are contiguous and act upon one another, the motion of one body is dependent on the motion of all the others, and therefore varies in numerous ways (Pr III 157)."[28] In a further analogy, this holistic emphasis is presented in a slightly more detailed manner. With respect to a circle of particles, with positions labeled EFGHI spaced along its circumference, he states: "If the air at E is pushed towards F, the air at F will circulate in the direction of GHI and return to E, such that one cannot feel its weight, in the same way that one cannot feel the weight of a rotating wheel if it is balanced perfectly on its axle (2 June 1631, AT X 205—see, Figure 19)." To envision the motion of a series of plenum particles as involving so little effort, i.e., as if they had no weight, strongly attests to the non-local synchronization or harmony of Cartesian motion (as well as to the continuing influence of simple machines on Descartes' thinking).

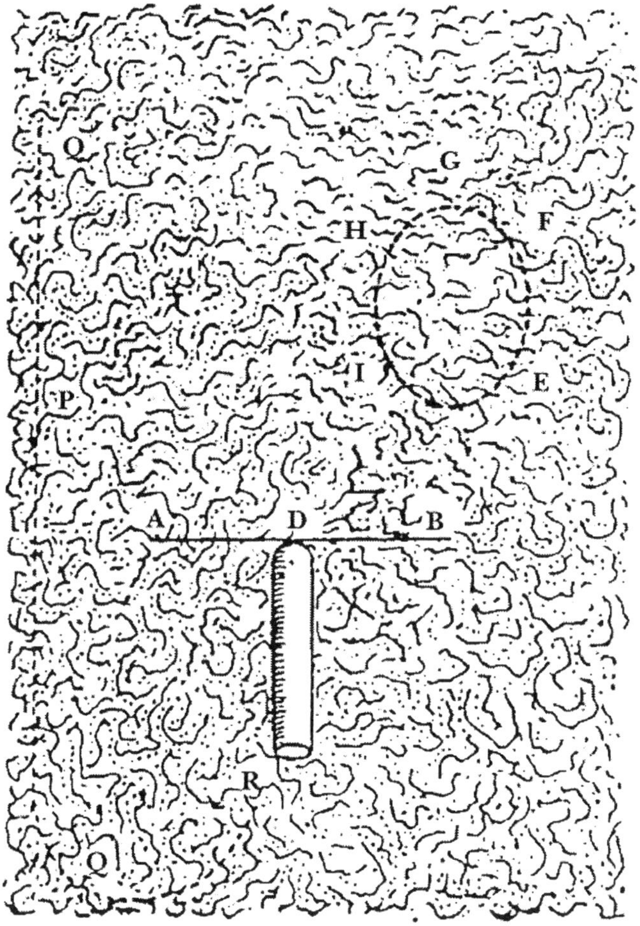

Figure 19. Descartes' diagram (from AT X 205).

This global, as opposed to local, feature of Cartesian natural philosophy is seldom discussed in the literature, although the work of a hand-full of commentators has investigated holistic notions as they pertain to the issue of teleology in general (but not specifically to the problem of QM as addressed here).[29] More typical of the attitude of commentators is a reductionist reading, as powerfully presented by E. Grosholz in commenting upon Cartesian matter and motion: "the whole is not greater than the sum of its parts, and there is no systematic interdependency among the parts, or among the parts and the whole (ibid., 76)." C. Merchant takes a similar line, contrasting the reductionist particle-based approach of the mechanical school with what she takes to be the anti-mechanist, "holistic" outlook of modern ecology. As for the latter, "no element of an interlocking cycle can be removed without the collapse of the cycle. . . . Each particular part is defined by and dependent on the total context. The cycle itself is a dynamic interactive relationship of all its parts, ..."[30] Ironically, this definition of a holistic system seems a more apt description of Descartes' cosmos than of our experience of ecological systems: e.g., the removal of a part of the Cartesian plenum would violate the conservation principle (and may derail the large scale circular motion of bodies, Pr II 33, or lock-up the harmoniously arranged vortex rotations, Pr III 65-66), whereas we know of many plant and animal species whose demise did not appreciably affect the larger ecosystem.

In all fairness, Grosholz' judgment is largely based on the vexed issue of individuation in the plenum (i.e., that bodies are individuated by motion, but motion is a relation among a bodily defined "place", Pr II 25), and Merchant does hint at, if only just (204), a possibly more complex interpretation of Descartes' brand of mechanism. These scholars have raised legitimate concerns, it should be acknowledged, since the potential circularity of Descartes' definitions, and the naive mechanism of some of his biological and physical models, constitute serious obstacles to the success of his program. But, Grosholz and Merchant extend the charge of crass reductionism to the entire Cartesian scheme of natural philosophy, and thus overlook the very tangible "systematic interdependency among the parts" revealed in the workings of QM. In the case of Merchant, for instance, some of the difficulties may lie in her characterization of the general features of Early Modern mechanistic thought, where it is alleged that "nature is made up of modular components or discrete parts connected in a causal nexus that transmitted motion in a temporal sequence from part to part (228)." This is not an accurate description of Descartes' *overall* theory, as should be evident from section 2, for Cartesian bodies interact simultaneously as a connected whole, and not "in a temporal sequence from part to part."[31]

5.3.3. Kinematics or Dynamics: The Letter to More.

Unfortunately, even if we acknowledge the global character of QM, this does not render the conservation principle any more hospitable to relational motion. The critic will surely respond that, even granting an entire synchronized plenum, QM must be determined by the motions of bodies—and motion is a relational change of place, thus raising once more the specter of arbitrary QM measurements (Part III will be largely devoted to attempts to resolve this problem). The instantaneous-tendency measure of QM may escape this dilemma, since as a "virtual" motion it presumably does not necessitate an actual change of place. Yet, the durational-"size × speed" version of QM clearly cannot fall back on a similar line of defense, which raises the difficulty first mentioned in section 1: How can motion be both a relational change of place and a conserved quantity? It would be presumptuous to declare that there exists an easy, if not definitive, answer to this major puzzle underlying Cartesian physics. Nonetheless, as argued above, the very question assumes that force, as measured by QM, can be entirely reduced to the sum of the size × speed of individual bodies—and, besides the arguments previously examined, there may exist textual evidence that contradicts this very assumption (at least in one notable, seldom discussed, instance).

In the late correspondence with More, Descartes' relational theory of motion faced one of its only serious documented trials. Provided that translation of a body is merely reciprocal (as change of place), More asked how one should regard the case of wind blowing through a tower window (5

March 1649, AT V 312): Is this motion reciprocal as well, such that the air can be viewed as resting and the tower in motion? Descartes' response reveals much about the interrelationship of kinematic and dynamic factors in Cartesian physics. Rather than answer More's question straightforwardly as a kinematics puzzle, he attempts to resolve the problem by resorting to a different, dynamics-based example. He asks us to picture two men attempting to free a grounded boat; with a man on board pushing against the shore, and the other on shore pushing against the boat:

> I cannot better explain the reciprocal force [*vires*] in the separation of two bodies one compared to the other than by putting under your eyes a situation wherein a boat is stranded against the shore of a river and there are two men, one of which is on the shore and pushes the boat with his hands away from the shore, and of which the other being in the boat, in the same way, pushes the shore with his hands to also draw the boat away from the shore. If the force [or strength (*vires*)] of the men is identical, the effort of the man on the shore, who is thus connected to the land, contributes no less to the boat's motion that the effort of the man on the boat, who is transported along with it. Therefore it is obvious that the action by which the boat recedes from the shore is equally in the shore as in the boat (15 April 1649, AT V 346).

Remarkably, Descartes' example employs a dynamics solution to an originally conceived kinematics problem: the reciprocity of the translation of the boat and shore, which is a rather embarrassing kinematical ramification of his espoused relationalism, is reinterpreted as a reciprocity of force between two possible dynamic sources of the motion, namely the push of each man.

One might view this "boat" example, as we will call it, as a sort of unintended conflation of the kinematic and dynamic elements of his theory, which is the suggestion offered by Shea (1991, 323). While this may be true, if we take the passage at face value (i.e., literally construed), then the very kinematic/dynamic distinction in Cartesian physics is seen in a different light, especially as it pertains to relational motion. The reciprocity "process" now appears to be directed at QM, as the measure of force, as much as it is directed at the kinematics of transfer. This realization has important consequences for our investigation, since it demonstrates that the force of motion *cannot be simply reduced* to the kinematical size × speed of individually moving bodies. At the very least, Descartes' "boat" example compels us to consider the force of motion, and thus QM, as ontologically on a par with the kinematics of motion. QM can be measured by size × speed in certain situations, as in the durational interactions of bodies, but QM is far too complex a notion to be fully encompassed as *just* the product of size and speed. Put simply, to regard QM as the mere byproduct of the reciprocity of translation is to envision Cartesian (relational) motion as somehow ontologically/epistemologically complete in itself (or self-evident, prior, etc.)—but Descartes' appeal to the dynamical factor in his "boat" example (to explicate his kinematics) undermines the plausibility of this hierarchical ordering.[32] In chapter 6, we will explore the possibility of utilizing Descartes' "boat" example as a means of constructing a relationally compatible system of reference frames for measuring QM, and thus try to resolve the inconsistencies in Cartesian relationalism (as raised, for instance, in Newton's argument).

5.4. Conclusions

As first mentioned in chapter 3, Descartes' conserved force of motion, QM, can be seen to harbor an intractable "qualitative" bias, which contrasts with the "quantitative" connotations that accrue from his formula for measuring QM, namely size × speed. Like many other Cartesian concepts, the conservation principle bears the vestigial imprint (if only just barely) of an unwittingly assimilated Scholastic outlook. The qualitative aspects of Descartes' conservation principle have been acknowledged before, principally as regards God's sustaining role (see footnote 30), but it has been the goal of this investigation to disclose some of these details as they pertain to the actual conserved value, QM, and its function in the Cartesian plenum. Not only is the conservation principle more complex and potentially resilient than some of its critics have charged (such as Nelson), but the traditional interpretation of QM as the product of size and durational speed misses much of functional scope of the concept. The instantaneous interconnected-ness of the Cartesian plenum, much like the simple machines his conservation principle may have been modeled upon, defies a classification purely along the lines of "size × speed". The letter to More would likewise appear to rule out any easy reduction of Cartesian dynamics to the kinematics of reciprocal change of place. As for its drawbacks, on the other hand, it would seem that the very Cartesian successes disclosed in our investigation have come at the expense of a conflating of the kinematic and dynamic aspects of Descartes' natural philosophy. And, indeed, a devoted Cartesian may wonder if a literal interpretation of the More letter is too high a price to pay for an effective means of handling the relational motion quandary.

In fact, as hinted above, there is a strange sense in which Descartes' conception and application of his conservation principle resurrects a sort of truncated version of a Scholastic substantial form. Despite his often repeated aversion to positing "mentalistic" substantial forms to *individual* bodies in order to explain their behavior, such as weight (e.g., *Sixth Replies,* AT VII 441-442), Descartes nevertheless invokes a force in the *whole of matter* that possess many of the characteristics of a substantial form: first, it is a quasi-"property" coupled (by God) to a pre-existing material substratum, the world (in this case); second, its teleological-like conservation of QM is irreducible to matter and motion alone (since God's sustenance is required, not to mention the import of the More letter). This "substantial forms" analogy should not be taken to extremes, however, for Descartes categorically rejected any such ontology. In his final letter to More, for instance, he admits that he had been previously disinclined to discuss his views on force out of a "fear of appearing disposed to favor the beliefs of those who regard God as a world-soul conjoined to matter (July 1649, AT V 404)." QM may not be God's soul united with the material world, yet many of the holistic, irreducible, and "mysterious" features of QM nevertheless betray a strong teleological predilection.

ENDNOTES

[1] C. Truesdell, "Experience, Theory, and Experiment", in *An Idiot's Fugitive Essays on Science* (New York: Springer Verlag, 1984), 6. Truesdell judges Descartes' physics to be a "failed attempt", however.

[2] Among the most notable of these early critiques, and portentous for the development of science, are the works by Huygens (a Cartesian, of sorts), Newton, and Leibniz (both adamantly non-Cartesian). See, C. Huygens, *De motu corporum ex percussione,* in *Oeuvres Complètes,* vol. 16 (La Haye: Société Hollandaise des Sciences, 1950); G. W. Leibniz, "Critical Thoughts on the General Part of the Principles of Descartes," in *Leibniz: Philosophical Papers and Letters,* ed. by L. E. Loemker (Dordrecht: D. Reidel, 1969); I. Newton, *De Gravitatione et aequipondio fluidorum,* in *Unpublished Scientific Papers of Isaac Newton,* trans. and ed. by A. R. Hall and M. B. Hall (Cambridge: Cambridge University Press, 1962). These works also address one of the other frequent objections to Descartes' conservation principle: namely, although it is true that there exists an invariant universal motive force or measure of motion, quantity of motion is not it. This topic will not be a primary concern of this essay, since it is a rather large and separate problem.

[3] A. Nelson, "Micro-Chaos and Idealization in Cartesian Physics", *Philosophical Studies,* 77, 1995, 377-391.

[4] R. J. Blackwell, "Descartes' Laws of Motion", *Isis,* 57, 1966, 222.

[5] Descartes does refer to a "quantity of motion" in this section from *The World* (*quantité de mouvemens,* which is strangely plural), but, once again, it is unclear if size × speed is what is meant.

[6] See, I. Beeckman, *Journal Tenu par Isaac Beeckman,* 4 vols. (The Hague: M. Nighoff, 1939-53), vol. III, 129. Although these statements of a conserved quantity date from the late 1630's, he probably had discussed such concepts with Descartes at a much earlier time. In addition, Beeckman's conservation principle is unusual in that he thinks that the impact of his hard bodies would ultimately result in a loss of motion if it were not for the infinite amount of motion God placed in the universe at its creation (which continuously "recharges" the bodies to maintain the same amount). On this theory, see A. Gabbey, "Essay review of W. L. Scott's, *The Conflict Between Atomism and Conservation Theory: 1644-1860,*" *Studies in History and Philosophy of Science,* 3, 1973, 373-385.

[7] See, P. Damerow et al., *Exploring the Limits of Preclassical Mechanics* (New York: Springer-Verlag, 1992), 103-126, and D. Garber, *Descartes' Metaphysical Physics* (Chicago: University of Chicago Press, 1992), 188-193, on Descartes' concept of determination and its status as a mode of corporeal substance.

[8] S. Nadler, "Deduction, Confirmation, and the Laws of Nature in Descartes' *Principia Philosophiae*", *Journal of the History of Philosophy,* 28, 1990, 366 (Nadler's emphasis). We will not enter into the complex debate on the alleged "necessity" of the laws of nature, i.e., how they resemble, if at all, the other so-called necessary truths of the Cartesian scheme, such as geometry or substance. For more on this complex problem, as well as on the relationship between the conservation principle and a priori truths based on God's immutability, see: C. Normore, "The Necessity in Deduction: Cartesian Inference and its Medieval Background", *Synthese,* 96, 1993, 437-454; B. Dutton, "Indifference, Necessity, and Descartes' Derivation of the Laws of Motion", *Journal of the History of Philosophy,* 34, 1996, 193-212; and, J. Broughton, "Necessity and Physical Laws in Descartes' Philosophy", *Pacific Philosophical Quarterly,* 68, 1987, 205-221. These studies reach similar inconclusive or negative results on the potential a priori grounding of QM.

[9] It us unclear, however, if Buridan utilized the *product* of velocity and quantity of matter for his measure of impetus. See, e.g., M. Clagett, *The Science of Mechanics in Middle Ages* (Madison: University of Wisconsin Press, 1959), chap. 8. Also on impetus theory, see A. Maier, *On the Threshold of Exact Science,* trans. by S. D. Sargent (Philadelphia: U. of Pennsylvania Press, 1982), chap. 4.

[10] Descartes' treatment lacks the mathematical concepts of the calculus which Jean Bernoulli employed in formulating the first "official" virtual work principle, or "virtual speed," as he called it. Here, Descartes' use of minute displacements almost certainly drew on the similar, earlier work of Guido Ubaldo, for he refers to Ubaldo in the October 11 1638 correspondence to Mersenne (AT II 380-399). For more on Ubaldo and Bernoulli, see, R. Dugas, *A History of Mechanics,* trans. by J. R. Maddux (New York: Dover Publications, 1988), 100-101, and 231-233. On statics in general, see, J. E. Brown, "The Science of Weights", in *Science in the Middle Ages,* ed. by D. C. Lindberg (Chicago: University of Chicago Press, 1978), 179-206. Descartes' QM might have also been influenced by Galileo's concepts of *impeto* and *momento,* since he was generally familiar with Galileo's work; e.g., the letter to Mersenne, 11 October 1638, AT II 379-419. On Galileo's and Descartes' work on statics, see, R. S. Westfall, *The Concept of Force in Newton's Physics* (London: MacDonald, 1971), chaps. 1 & 2. Gabbey notes, 1973, 383, that "quantity of motion," size × speed, may have been a fairly common measure of motion among seventeenth century natural philosophers.

[11] A. Gabbey, "Descartes's Physics and Descartes's Mechanics: Chicken and Egg", in *Essays on the Philosophy and Science of René Descartes,* ed. by S. Voss (Oxford: Oxford University Press, 1993).

[12] The inconclusive debate on Descartes' understanding of instants, i.e., whether instants are durationless or comprise some length, will not be discussed (since it is a separate problem). See, Garber, 1992, 268-270, for a discussion of the respective sides in this controversy; and, D. Des Chene, *Physiologia: Natural Philosophy in Late Aristotelian and Cartesian Thought* (Ithaca: Cornell University Press, 1996), 280, fn. 33. As used in this essay, "instant" and "instantaneous" will allow for either reading, just as long as the corresponding concept of 'beginning to fall' mentioned in his statics correspondence (AT II 229-233) is likewise allowed both interpretations (in order to secure the arguments put forth in this section).

[13] Consequently, determination applies to both instantaneous tendencies and non-instantaneous motions (as Garber likewise concludes, 1992, 219-220). It is important to keep these two notions separate, for a misrepresentation of Descartes' views can result if this difference is not heeded. For example, Demarow et al., 1992, 77, fn. 15, appear to believe that determinations are only instantaneous quantities, and QM non-instantaneous, based on the oversight of this distinction, as well as on an oversight of the overall role of tendencies in Cartesian physics.

[14] As Westfall, 1972, 61-62, has pointed out, Descartes' use of the term 'agitation' is somewhat ambiguous. Yet, he agrees that it signifies 'momentum' (or QM, if it is a scalar property) in the articles on stellar motions in Pr III of the *Principles*. E. J. Aiton also reaches the same conclusion; see, *The Vortex Theory of Planetary Motions* (London: MacDonald, 1972), 63, fn. 83. As regards QM's alternative role as a gauge of instantaneous bodily tendencies, both Garber (1992, 208) and T. Prendergast concede this basic point along lines similar to those advanced above. See, T. Prendergast, "Motion, Action, and Tendency in Descartes' Physics", *Journal of the History of Philosophy,* 13, 1975, 453-462.

[15] For Descartes on the concept of infinitesimals, see, P. Mancosu, *Philosophy of Mathematics and Mathematical Practice in the Seventeenth Century* (Oxford: Oxford University Press, 1996), 82-82, 142-143. Mancuso argues convincingly that Descartes attitude towards infinitesimals is more ambiguous than historians have hitherto acknowledged, since he at times made use of them: see the letter to Clerselier, July 1646, AT IV 443.

[16] Of course, Descartes' natural laws do not constitute the present-day concept of inertial motion, but his theory is a close forerunner to this notion. See, e.g., A. Gabbey, "Force and Inertia in the Seventeenth Century: Descartes and Newton", in S. Gaukroger, ed., *Descartes: Philosophy,*

Mathematics, and Physics (Sussex: Harvester Press, 1980), 230-320. See, Pr IV 20-27, for Descartes' analysis of terrestrial gravity.

[17] See, S. Gaukroger, *Descartes: An Intellectual Biography* (Oxford: Oxford University Press, 1995), 247.

[18] See, W. R. Shea, *The Magic of Numbers and Motion: The Scientific Career of René Descartes* (Canton, Mass.: Science History Publications, 1991), 27-30. In a later work, Gaukroger has argued quite persuasively for the conflation, in Descartes' natural laws, of a kinematically-conceived concept of rectilinear motion and a statically-conceived concept of bodies moving in circular equilibrium. The unwitting conjunction of these two disparate approaches to the phenomena of bodily motion is largely responsible for the "two aspects" of the Cartesian conservation law, as described below (although Gaukroger does not explore in depth the instantaneous/non-instantaneous dual aspect of QM). See, S. Gaukroger, "The Foundational Role of Hydrostatics in Descartes' Natural Philosophy", in *Descartes' Natural Philosophy,* ed. by S. Gaukroger, S. Schuster, and S. Sutton (London: Routledge, 2000).

[19] See also the letter to Constantin Huygens, March 9 1638, AT II 662. On the topic of "dimensions", see, J. A. Schuster, "Descartes' *Mathesis Universalis: 1619-28,*" in S. Gaukroger, 1980, 64-69.

[20] One again, the status of Descartes' natural laws as regards their "necessity" is much debated: see references listed in footnote 8 for a discussion of the eternal truths and their relationship to necessity in Descartes' physics. Gabbey's use of Descartes' "possible worlds" description of (some form of) necessity stems from *The World,* AT XI 47.

[21] The preceding analysis may be reading too much into Gabbey's assertions, however. He may be implying that three-dimensional gravity and two-dimensional statics are *both* equally contingent aspects of this particular world (since a different vortex system could be imagined). The tenor of his argument, nonetheless, seems to confer, at least to me, a higher level of necessity on three-dimensional gravity than on two-dimensional statics, which is the basis of the disagreement.

[22] For example: "On earth, we see that gold, lead, or other metals when moved retain more agitation, or force to continue in their motion, than do pieces of wood or rocks that have the same size and shape; and therefore metals are also believed to be more solid [or hard, *solidus*], or to contain more third element matter and have smaller pores filled with first and second element matter (Pr III 122)." For more on Descartes' "indefinite" notion, see, J. E. MacGuire, "Space, Geometrical Objects and Infinity: Newton and Descartes on Extension", in *Nature Mathematized: Historical and Philosophical Case Studies in Classical Modern Natural Philosophy,* ed. by W. R. Shea (Dordrecht: D. Reidel, 1983).

[23] G. E. Mase, *Theory and Problems of Continuum Mechanics* (New York: McGraw-Hill, 1970), 44. For more on the idealizations in continuum mechanics, see, M. Wilson, "There's a Hole and a Bucket, Dear Leibniz", in P. A. French, T. E. Uehling, Jr., H. K. Wettstein, eds., *Midwest Studies in Philosophy Vol. XVIII, Philosophy of Science* (Notre Dame, IN: U. of Notre Dame Press, 1993). It should be noted, however, that Wilson's position on Cartesian physics differs considerably from the views expressed in this essay.

[24] See, A. Gabbey, "The Mechanical Philosophy and its Problems: Mechanical Explanations, Impenetrability, and Perpetual Motion", in *Change and Progress in Modern Science,* ed. by J. C. Pitt (Dordrecht: D. Reidel, 1985), 57; and, E. Grosholz, *Cartesian Method and the Problem of Reduction* (Oxford: Clarendon Press, 1991), 90. One of the earliest versions of this argument appears in Book II of Newton's *Principia.* See, I. Newton, *Mathematical Principals of Natural Philosophy,* trans. by A. Motte and F. Cajori (Berkeley: University of California Press, 1962), 391.

[25] It is assumed, here, that the plenum as a whole does not have a "place" (contiguous neighborhood) from which it could move relative to, and thus conserve QM. Many of the Scholastics would have also rejected this possibility, for the place of an indefinitely extended plenum would seem a conceptual contradiction. For more on the historical development of these theories and arguments, including Leibniz's, see, E. Grant, *Much Ado About Nothing: Theories of Space and Vacuum from the Middle Ages to the Scientific Revolution* (Cambridge: Cambridge University Press, 1981).

[26] Although it is beyond the bounds of this essay, some commentators would contend that Descartes' theory does not sanction the sort of "reference frame" relationalism seemingly implied in Westfall's allegations: see, D. Garber, 1992, 169-171, and D. Des Chene, 1996, 270-271. Garber's and Des Chene's arguments are mainly correct, but they still do not free Descartes' natural laws from relational contradictions; see chapter 6.

[27] Once again, "kinematic" refers to the study of motion in the absence of forces, unlike "dynamics", which does study motion under the action of forces. In addition, "invariant", as used in this context, can mean either; (a) a numerical value which remains the same from the perspective of all frames, but is not necessarily the same numerical value in each frame; or (b), the same as (a), but the numerical value *is* the same in all frames. The type of invariant in classical conservation laws, such as Newton's momentum law, is (a); while the invariant spacetime interval of Special Relativity is (b). See, M. Friedman, *Foundations of Space-Time Theories* (Princeton: Princeton University Press, 1983), 56, for more on spacetime invariants. Descartes' QM violates either reading, if relationalism is also in play. For example, if in one frame of reference, a small moving body merely reverses its motion (from right to left) after striking a larger resting body, then QM will be

conserved as mandated by rule four (Pr II 49; and where the scalar nature of speed is in evidence). But this same collision as viewed from another frame, at rest relative to the smaller moving body, will see both bodies retreat to the left after impact (i.e., this frame, assuming it is inertial, will keep moving to the right)—and this joint motion of both bodies will (in most cases) register an *increase* in QM *after* the collision (thus violating the conservation principle).

[28] Descartes' hypothesis of instantaneous circular motions of plenum bodies (Pr II 33) is also relevant to this discussion, needless to say, although it is clear from the context that *circular* motions are necessary, at least in Pr II, to avoid the creation of a vacuum. The interconnected-ness of *all types* of motions, however, seems to stem from other sources, as argued above.

[29] For some studies of teleology in Cartesian physics, see, J. Collins, *Descartes' Philosophy of Nature* (Oxford: Blackwell, 1971), Part 1; and, P. K. Machamer, "Causality and Explanation in Descartes' Natural Philosophy", in *Motion and Time, Space and Matter: Interrelations in the History and Philosophy of Science,* ed. by P. K. Machamer and R. G. Turnbull (Columbus: Ohio State University Press, 1976).

[30] C. Merchant, *The Death of Nature: Women, Ecology, and the Scientific Revolution.* (New York: HarperCollins Pub., 1980), 293.

[31] Merchant also errs, it would seem, in reading overt mechanist overtones in the *Discourse on the method* (1980, 231). From the context of the discussion, however (AT VI 19-20), the a priori certainty and structure of geometry is the more obvious analogy that Descartes is attempting to draw in formulating his methodological rules.

[32] Finally, it should be mentioned that the conclusions reached in this essay appear to be neutral with respect to the debate over the ontological status of Cartesian force (see chapter 3).

PART III

CONSTRUCTING A CARTESIAN SPACETIME

CHAPTER 6

RELATIONAL SPACETIMES AND CARTESIAN DYNAMICS

Part III begins the lengthy examination of possible Cartesian spacetimes that can accommodate Descartes' problematic relational theory of motion, especially when viewed within the context of his dynamics, the Cartesian natural laws. Whereas Part I dealt only with the general structure or features that a Cartesian spacetime must possess, that is, if the spacetime is to be considered a truly relational spacetime, the specific details of such spacetimes were left undeveloped. Starting in chapter 6, consequently, and extending through chapter 10, we will now attempt to specify the particulars of a working, consistent Cartesian spacetime. As will be evident throughout our investigation, the diverse hypotheses and concepts which comprise Descartes' physics, as explored in Part II, will have a major role to play in guiding the construction of these possible spacetimes. Chapter 6 begins this project by overviewing the work of various commentators on the interpretation of Cartesian motion, with particular emphasis placed upon the recent important texts of D. Garber and D. Des Chene. In contrast to the methodology of most previous treatments of Cartesian motion, however, Part III, like Part I, employs a modern "spacetime" approach to the problem. By this means, chapter 6 will focus on the relatively neglected role of Cartesian dynamics, as opposed to kinematics, in attempting to find a solution to the puzzle of Cartesian motion (as was introduced in chapter 5).

Specifically, our modern approach to the problem of Cartesian motion will not only aid in the understanding Descartes' complex theory, but it will also reveal the shortcomings of many previously attempted solutions, including both Garber's and Des Chene's (sections 3 and 4). In the final section of the chapter (section 5), a deeper, underlying reason for the limited success of these earlier readings will be disclosed: namely, commentators have neglected the dynamics of Descartes' system (that pertains to bodily motions under the actions of forces) in favor of the kinematics (that concerns bodily motion per se). Put differently, whereas most commentators have attempted to treat the kinematical problems of Cartesian physics before dealing with its numerous dynamical inconsistencies, a more serviceable solution to the problem of Descartes' kinematics might take the dynamics as primary, and thus reverse the course of the standard interpretation. As will be demonstrated, the (dynamic) Cartesian collision rules can play an essential role in picking out the privileged reference frames needed to secure a consistent relational account of (kinematic) Cartesian motion.

6.1. Cartesian Space and Motion.

Up to this point in our investigation, we have been content to treat Descartes' hypotheses on motion, as well as space, as a species of "relationalism"; i.e., which rejects that space and motion are, or require, anything over and above the relations among bodies. From Descartes' time down to our own, his advocacy of a relational hypothesis has been taken for granted, and has prompted a number of negative critical responses and lengthy denunciations, both from the "absolutists" who deny relationalism, such as Newton, and from fellow relationalists committed to his general, if not exact, goals, such as Leibniz. Most of these difficulties, of course, center upon the incompatibility of Descartes' espoused relationalism with his proto-Newtonian series of natural laws. Recently, however, some commentators, notably Daniel Garber and Dennis Des Chene, have presented strong arguments purporting to show that Descartes' theory of motion has been largely misunderstood by his critics. Descartes did not advocate a strong form of relational motion in the modern sense, they claim, and any attempt to read such a theory into his natural philosophy misconstrues the historical context in which it was developed.

Although it is sensible to be wary of the implicit dangers of over relying on modern notions, an interesting, if unintentional, question is thus raised: If Descartes' theory has been mistaken for a crude form of early relationalism, then which type of relationalism was it assumed to be? As John Earman has persuasively argued (1989), there are several discernible variations on a relationalist theme, thus it seems legitimate to inquire into the precise details of the *type* of relational theory of motion, and space, that Descartes' allegedly employed.

6.1.1 Cartesian Space.

As has been well documented, the vexed issue of the nature of space and time that Descartes inherited from the Medieval/Renaissance thinkers had a long and venerable tradition (see, Grant 1981; although how much influence Descartes' Jesuit training had on his mature theory of space remains unclear; Garber, 128). While some philosophers had begun to question the Aristotelian dictum that whatever had dimensionality was body, and to envision space as some form of substantial entity, the paths they followed from this point diverged considerably. On the one hand, philosophers such as Telesio, Campanella, and Bruno, held space to be always filled with matter yet independent of it; while others, like Patrizi, and Descartes' atomist contemporary, Gassendi, accepted this "container" view but endorsed empty spaces totally devoid of matter.

Descartes' hypotheses on space ran against the grain of this anti-Aristotelianism, however, for he unequivocally rejected the existence of a vacuum and of a space separate from body (Pr II 16). Indeed, recalling our discussion from Part II, one of the most well-known principles of Cartesian metaphysics is his "identification" of material body and spatial extension: "the extension in length, breadth, and depth which constitutes the space occupied by a body, is exactly the same as that which constitutes the body (Pr II 10)." This novel solution thus allowed Descartes to take a nominalist stance on the concept of "space," which he also dubs "internal place," since

> we attribute a generic unity to the extension of the space [of a body], so that when the body which fills the space has been changed, the extension of the space itself is not considered to have been changed or transported but to remain one and the same; as long as it remains of the same size and shape and maintains the same situation [*situm*] among certain external bodies by means of which we specify that space (Fr Pr II 10).

Consequently, by employing a set of external bodies as a reference frame to delimit the general extension of a part of the plenum (i.e., a world completely packed with matter), one can secure an "abstract" notion of space. Descartes takes a similar view of "time", which is a generalized abstraction from the "durations" of particular objects (where duration is an attribute of substance; Pr II 56-57). In addition, Descartes' understanding of space, as the quotations above confirm, is three dimensional, but whether or not he viewed the geometry of space as Euclidean is not specified. Patrizi and Newton seem to be the only natural philosophers of this general period who explicitly ascribed this property to space (Grant, 232-234). Given the lack of any known alternatives, however, Descartes probably accepted this view, as well (and we shall assume this below).

With regard to the actual "ontology" of space, J. Bennett has pointed out that the precise nature of Descartes' views are somewhat obscure.[1] The denial of a vacuum, which he envisions as a sort of "extended nothing" (i.e., extension as a property of nothing), follows directly from his definition/identification of matter as spatial extension. Yet, simply ruling out the possibility of a vacuum does not necessarily entail a fully developed relational theory of space, for one can retain some primitive notion of space as a "container," or as some process of bodily identification which grounds spatiotemporal individuation, while continuing to deny the possibility of a vacuum. Descartes, however, remains silent on these alternative formulations of non-relational spacetime. Given his analysis of the "empty vessel", as discussed in chapter 3, nevertheless, it seems safe to conclude that Descartes would have not been cordial to any weakened formulation of an absolute spacetime that can exist *apart* from material bodies (see section 6.3 below, for more on this topic). In the case of the vessel, he reasons:

> If someone were to inquire as to what would occur if God removed the whole body contained in a vessel and did not allow anything to take the place of the body that had been removed, the answer must be that the sides of vessel would thus become contiguous to one another. For, when there is nothing between two bodies, they must necessarily touch one another; . . . (Pr II 18).

According to Descartes, therefore, removing the matter within a vessel also removes its extension, with the inevitable outcome that the sides would now be immediately contiguous; i.e., the sides of the vessel do not move into an empty space that exists between their former positions (as is revealed more explicitly in the correspondences that accompany this thought experiment; AT IV 109, AT II 482). In short, without matter, there simply is no space/extension, a conclusion which seems to exclude the existence of some lesser form of absolute space (which could provide the ground for the spatiotemporal properties of individual bodies, etc.).

Bennett also seriously considers the notion of, what he calls, a "separator" space, which can be envisioned as analogous to a fish swimming in water: "There are regions of space and there are bodies, and the two kinds jointly exhaust all the locations there are; but no two particulars have the same position at the same time. Where body starts, space stops, just as where fish starts, water stops (13)." On a container space reading of this example, it would be possible to talk about the fish and the water collocated in the same spot, but not on the separator space interpretation, of course. Bennett eventually concludes that this formulation of a Cartesian space does not work, because the minute material particles (first degree matter) in Descartes' plenum are not entirely free of a force of resistance; i.e., even minute particles resist the motion of larger moving bodies—and they would need to be nonresistant if it were a true separator space (21-23). While Bennett is correct in concluding that all matter, even primary, possess some degree of resistance (in the interactions with other bodies), his approach to the problem of bodily individuation seems to presuppose that there is an ontological distinction between what counts as a body and what counts as subtle matter (primary matter for the latter, and secondary and higher forms of matter for the former). Yet, since all matter is the same for Descartes, being demarcated only by relative motion, any attempt to ground separator space on the size, and/or motion, of particles seems to presuppose motion, and hence space (since "place," which is the means of determining relative motion, is derivative of space). This circularity is duly noted by Bennett, it should be added (21-23). If one leaves out Bennett's hypothesis on resistance, and on some possible "real" distinction between bodies and subtle matter, then all that remains of his hypothesis is that the material occupants of Descartes' plenum cannot be co-located. The impossibility of co-location is clearly an accepted feature of Descartes' view (especially when the letter to More, AT V 346, is taken into account). Descartes' working hypothesis concerning space, consequently, is the rather "minimalist" notion that all parts of material extension, no matter what their size, shape, etc., are separated from all other material parts of extension (at any given moment, since bodies change)—and it is not clear under what category such a meager view space can be classified.

6.1.2. Cartesian Motion.

That Descartes' analysis of motion has been regarded as rather blatantly relationalist is understandable given Pr II of the *Principles of Philosophy,* the work which presents his most extensive discussion of this phenomena. Once more, it is in this work that Descartes puts forth his principle definition of motion (*motus*) as "*the transfer [translatio] of one piece of matter or of one body, from the neighborhood [vicinia] of those bodies immediately contiguous to it and considered at rest, into the neighborhood of others* (Pr II 25)." As noted by most commentators, Descartes deliberately wants to distinguish his "proper" conception of motion as change of "neighborhood" from the common or "vulgar" conception of motion as a change of "place" (*locus*); where, as above, "(internal) place" is defined as the situation of the body relative to some arbitrary set of, usually resting, distant bodies (see, Pr II 10-15, 24-28)." Descartes employs these definitional distinctions in a decidedly baroque manner, however, for he also characterizes "external place" as the surface of the containing bodies, thereby rendering this new idea practically equivalent to his neighborhood concept (as Garber also concedes, 346, fn. 11). Regardless of these definitional puzzles, relationalism seems to arise in the following manner (see, also, chapter 1): in Descartes' plenum universe, any attempt to regard the surrounding bodies "at rest" amounts to a mere stipulation, since "we cannot conceive of the body AB being transported from the vicinity of the body CD without also understanding that the body CD is transported from the vicinity of the body AB (Pr II 29)." Hence, "all the real and positive properties which are in moving bodies, and by virtue of which we say they move, are also found in those [bodies] contiguous to them, even though we consider the second group to be at rest (Pr II 30)."

Following Earman's useful classificatory scheme (12), Descartes' hypotheses seem to advocate the following relationalist conception of motion:

R1: All motion is the relative motion of bodies, and consequently, spacetime does not have, and cannot have, structures that support "absolute" quantities of motion.

Motion, as conceived according to (R1), is relational in the strict sense, since Descartes seems to assert in his phrase, "considered at rest" (*tanquam quiescentia spectantur*), that the choice of which bodies are at rest or in motion is purely arbitrary, or relative to different frames of reference. Hence, since different perspectives will assign conflicting values to the very same body, there can be no "actual" or "absolute" determinations of an individual body's state of motion (see chapter 1). Most commentators regard (R1) as Descartes' preferred hypothesis of motion, albeit with definite qualms about his consistent application of it (Dugas, 178-179, Westfall, 57-58, Shea, 322-323, Earman, 41, Barbour, 449-450, to name only a few), while others have suggested that Descartes' avowal of (R1) is, at least in part, a sop to church censorship in order to advance his Copernican brand of planetary vortex theory (e.g., Koyré, 265, Blackwell, 227). [2]Garber, following Henry More, aptly dubs this (R1) view of motion, the "reciprocity of transfer" (167), and it is not difficult to comprehend why it has generated such controversy over the centuries. Not only does this hypothesis appear to run afoul of the vary criticisms Descartes leveled at the "vulgar" conception of motion, i.e., rendering motion hopelessly relative to conflicting perspectives (see, Pr II 24-28), but (R1) motion is also inconsistent with the laws of motion he advanced in the *Principles* (as disclosed more fully in chapter 3).

As first presented in Part I, problems begin to arise for Descartes when we conjoin his analysis of motion with his dynamics of bodily interactions. In the *Principles,* for instance, Descartes advocates a series of natural laws that appear to flagrantly violate relationalism by invoking determinate individual bodily states of motion: "all movement is, of itself, along straight lines. . . (Pr II 39)." Unfortunately, it is not possible, or meaningful, to attempt to ascertain the "unique" path or trajectory of a body given (R1). Since trajectories are determined relative to each observer, and all observers are in relative motion, any effort to fix the unique path of a particular moving body will result in a host of conflicting measurements, none of which can lay claim to its "actual" path.

Furthermore, the *Principles* betrays the influence of a second non-relational factor in Descartes' rendition of bodily "modes"; where, briefly, a mode is a particular instantiation or "way" that a corporeal body manifests its spatial extension (as with "shape": see, Pr I 56, 61). Some commentators, like Gaukroger (371-377), have rejected Descartes' apparent sanction of (R1) based on his insistence that "movement and rest are merely two diverse modes of [a] body (Pr II 27)." According to Garber, on the other hand, motion as a mode need not conflict with the reciprocity of transfer: "Unlike shape, motion seems to be relational; though there may be a genuine distinction between motion and rest, motion seems to be a property that pertains not to an individual, but, in a strange way, to both an individual and its surrounding neighborhood (172; these issues will examined further below)." Even granting Garber's point, a similar distinction between motion and rest is drawn in the first law of nature, where both are submitted as intrinsic, but different, natural bodily states: bodies do not tend towards rest, he reasons, since "rest is the opposite of movement, and nothing moves by virtue of its own nature towards its opposite or its own destruction (Pr II 37)." Besides revealing the influence of the Aristotelian/Scholastic logic of contrary predicates, Descartes' pronouncements run afoul of relationalist doctrine, for he seems to be presuming that rest and motion are not only distinct but opposing states of bodies, a qualitative difference that cannot be countenanced on any strict relational theory (see chapter 3).

6.2. Cartesian Relationalism.

In this section, we will investigate the problem of an (R1) construal of Cartesian motion, and thus set the stage for the introduction of alternative formulations of his relationalism.

6.2.1. Relational Spacetime.

The Cartesians who took up the cause of Descartes' natural philosophy, such as Régis and Rohault, mainly followed the guidelines of the theory of space and motion as laid down in the *Principles* (Clark, 126). Nevertheless, though they presented motion as a change of the contiguous bodily neighborhood, many of the Cartesians were wary of the apparent conflicts in Descartes' hypotheses. Rohault, to take one example, substantially simplified the collision rules in his rendition of Cartesian physics, since Descartes' harbor many anti-relational inconsistencies (as noted in chapter 3, Rohault, 48-50). Eventually, many natural philosophers, seeing that the obstacles were too great, began to tinker with the foundations of Descartes' relationalism in the hope of salvaging it; while others were lead to reject it outright.

Newton's early *De gravitatione* argument, as explored in chapter 2, is one of the best known historical instances of the latter approach, for it provides an extended critique of the inconsistencies in Descartes' theory, ultimately concluding that "it follows indubitably that Cartesian motion is not motion (Newton, 131)." Newton insists that "as it is impossible to pick out the place in which a motion began, for this place no longer exists after the motion is completed [due to the constant change of relative body position in Descartes' plenum], so the space passed over, having no beginning, can have no length; and hence, since velocity depends upon the distance passed over in a given time, it follows that the moving body can have no velocity [given the Cartesian definition of motion as change of (external) place] (131)." Although a more thorough examination of Newton's conception of space and time was provided in chapter 2, a few additional observations need to be drawn since they will play a factor in our later discussion. First, despite being only one of several arguments aimed at Descartes, Newton's *De gravitatione* argument (from chapter 2) is insightful in that it discloses the exacting Newtonian demand for a method of precisely tracking spatial positions across time. Although a relationally-conceived reference frame technique of determining motion might seem compatible with Newton's criterion (since the reference frame can continue to identify the same spatial positions relative to its perspective), the *Principia* makes it clear that only the "absolute" spatial positions of the Full-Newtonian models are sanctioned (being thus equipped with a spacetime "rigging" to identify such positions; Friedman 1983, 74): "Relative space is some movable dimension or measure of the [immovable] absolute spaces (Newton 1962b, 6)." Second, as presented in the *De gravitatione,* Newton followed Gassendi in carefully separating space and time from the categories of substance and property, for he reckons that space "has its own manner of existence that fits neither substances nor accidents (1962a, 132; see, McGuire 1978)." Yet, even if Newton's theory does fall short of positing space as a form of "substance," a view often dubbed "substantivalism, the *De gravitatione* tract also entertains the more radical notion, albeit tentatively, that space may be the only predicable (non-spiritual) substance, a thesis known as "supersubstantivalism". He imagines a scenario wherein God ascribes particular bodily properties to regions of space, such as impenetrability, and allows these collections of properties to migrate through space "according to certain laws, yet so that the amount and shape of that impenetrable space are not changed (1962a, 139)."

If Newton's supersubstantivalist musings in the *De gravitatione* are dismissed as a mere aberration, it may be a shock to discover that his views bear an uncanny resemblance to the spatiotemporal designs of some of the Cartesians, or at least to certain controversial interpretations of their views. We have already examined some of these modern interpretations of Cartesian space/matter (in chapter 4): namely, the "monist" theory of Cartesian material substance. While is unlikely that Descartes held such a view (see chapter 4), some of the Cartesians apparently did accept a form of monism with regard to matter; as J. Bennett's reading of Spinoza indicates: "If there is . . . a pebble in a region R, what makes this true is the fact that R is *pebbly,* where 'pebbly' stands for a certain monadic property that a spatial region can have (Bennett, 70-71)." T. Lennon, on the other hand, provides an interpretation of Descartes that places the individuating properties of bodies entirely within the mental realm, thus denying that there is any real motion or change in Descartes' one corporeal extension: Lennon claims that "motion ontologically is the flow of sensations", much like "a shape distinguished from the visual field by a difference in color will move when qualitatively the same shape is distinguished in an adjacent area of the field by the same color (Lennon, 204-205)." To repeat the criticism of Lennon supplied in chapter 4, Descartes constantly refers to motion as a mode of, or *in,* body/matter, and not as some sort of phenomenal division of a single material substance. As the following representative explanation demonstrates: "figure cannot be understood except in an extended thing, nor can motion, except in an extended space; nor can imagination, sensation, or will, except in a thinking substance (Pr I 53)."

Returning to our analysis, there have been many critics besides Newton who have charged Descartes' ostensible (R1) relationalism with undermining the very reality of motion. Leibniz argued that, "if there is nothing more in motion than this reciprocal change, it follows that there is no reason in nature to ascribe motion to one thing rather than to others. The consequence of this will be that there is no real motion (Leibniz, 393)." Likewise, T. Prendergast insists that "if we are to take this text seriously the reality of rest and motion is destroyed (Prendergast, 104)." Presumably, these commentators hold that if a property, such as rest or motion, cannot be posited unambiguously to a single body, then the "reality" of motion as a physical process is endangered. But, this form of reasoning misconstrues basic facts of relational spacetimes: for, although individual assignments of rest or motion to bodies are relative to a choice of perspective, this does not undermine the reality of motion if conceived as a relative difference in motion among bodies, that is, as "relative speed (velocity, acceleration, etc., see chapter 2)." According to (R1), only absolute quantities of motion, which are relative to a fixed, absolute spacetime structure, are ruled out; not the reality of motion itself.

More specifically, the Cartesian spacetime we are investigating forms a member of the larger class of Leibnizian spacetimes, which possess a Euclidean spatial metric on the planes of absolute simultaneity, or "time-slices" of the spacetime manifold, and also a time metric to uniquely order these time-slices (see chapter 2). The invariant quantities of this structure are, of course, the "relative speeds, etc."; and, as in our earlier discussion, these relative motions are ambiguous with regard to the assignment of *individual* component motions: if, for example, one calculates a speed difference of 10 knots among two bodies, the "actual" determination of each individual's state of rest or motion is just not possible—yet, the 10 knot "speed difference" between the bodies is assuredly real and will be agreed to by all. The primary reason for regarding Cartesian motion as favoring the Leibnizian structure, as opposed to the more sparse Machian spacetime (which lacks a time metric), is due to Descartes' penchant for utilizing a robust sense of the concept "speed" in the *Principles* (as will be discussed below). Relative speed is not an invariant of Machian spacetime and hence cannot be meaningfully ascertained; although if construed as a change in the instantaneous spatial separation among bodies (when compared across time-slices), then a very weak form of invariant motion could be tolerated by the Machian models (see Earman, chapter 2).

6.2.2. Garber and Des Chene on Relational Motion.

Based in part on their tacit understanding of Descartes' hypothesis of motion as a change of contiguous neighborhood (section 6.1), both Garber and Des Chene reach the identical conclusion that motion is an objective feature of the Cartesian plenum which cannot be simply transformed away by switching to a different reference frame. Using the terminology of "vulgar" motion or rest, which is measured relative to an arbitrarily assigned "(internal) place", and "proper" motion or rest as change of neighborhood, Garber asserts with respect to a motion between two bodies, A and B:

> Even though it is arbitrary whether or not B is at [vulgar]-rest, once we *consider* B at [vulgar]-rest, it is *not* a matter of arbitrary choice whether or not to consider A as being in [vulgar] motion; if A is *really* in motion in [Descartes'] proper sense of the term, if it is *really* separating from its neighborhood B, then no mere change of perspective will allow us to set A at rest in the *proper sense* (169).

Des Chene argues along similar lines, stating that "the nonrelational facts upon which judgments about motion are based are reciprocal facts of touching and not touching (271)," and thus "motion is always entirely actual, the instantaneous rupture of a body from its neighbors (256)." In expressing this objectivity of motion, however, Des Chene employs the notion of a "physical trajectory", which he defines as "a series of places, in the generic sense, which joined together constitute a tubelike virtual body (270)." That is, as a relative translational displacement occurs between a body A and its neighborhood B, a generic place is left behind such that, "if we imagine the trajectory to be generated by a succession of [these] translations, each of them adding to its increment, then the trajectory, like its parts, will be defined absolutely (270)." The complications inherent in this idea we will postpone discussing until section 6.3, but, for our present purposes, it is important to note that, like Garber, Des Chene's physical trajectory is an attempt to capture the objectivity, or "absoluteness", of motion in the Cartesian plenum even granting its reciprocal character.

The upshot of all this is that, having grasped onto an objective, non-relational facet of Cartesian motion, Garber and Des Chene believe they are now free to dismiss some of the long-standing doubts about the consistency of the Cartesian natural laws, especially the collision rules. As explained in chapter 3, there is a distinct non-relational undercurrent to Descartes' analysis of impact, as is most evident in the case of rules four and five: briefly, in the fourth rule, a large object remains at rest during impact with a smaller moving body, and simply deflects the smaller body back along its path (Pr II 49); whereas in the fifth rule, a large body will move a smaller stationary object, "transferring to [the smaller body] as much of its motion as would permit the two to travel subsequently at the same speed (Pr II 50)." From a relational standpoint, however, rules four and five constitute the same type of collision, since they both involve the interaction of a small and large body with the same relative motion (or speed difference) between them (see Figure 20). Therefore, because they represent an identical scenario, a consistent relational theory demands an identical outcome, which, unfortunately, Descartes does not provide. Garber renounces this conclusion, nevertheless:

> For Descartes, the case in which [a body] is in motion is physically distinct from the case in which it is at rest. And so, for him, the situations described in [rule four] and [rule five] are not mere redescriptions of one another; one cannot arbitrarily designate which of two bodies in relative motion is in motion and which is at rest (241).

Equipped with a means of discerning rest from motion—translation of neighborhood or no translation—the fourth and fifth rules can thus be individuated without transgressing relationalist tenets (Des Chene makes the same point, 297-298).

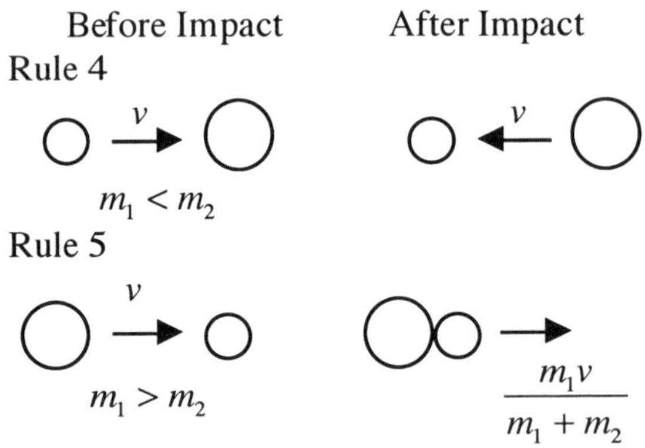

Figure 20. Descartes' Fourth and Fifth collision rules.

The problem with this line of reasoning, to put it bluntly, is that it only works if one presupposes that the two bodies are approaching one another, which is a property of the impact phenomenon not captured by sole reference to the neighborhood of each individual body. Even if there is relative motion between a body and its neighborhood of contiguous bodies, which we will dub, "local translation", it is still not possible to determine which collision rule the impact will fall under, or if the bodies will even collide at all, unless some reference frame is referred to that can compute the *motion of both bodies relative to one another*. Suppose, for instance, that a spatial interval separates two bodies of which one is, and the other is not, in local translation. Given this scenario, it is not possible to determine if; (i) the "translating" body is approaching the "non-translating" body, or (ii) the spatial interval between them remains fixed and the translating body simply undergoes a change of neighborhood (i.e., the neighborhood moves relative to a stationary body—see Figure 21). Consequently, Garber's and Des Chene's appeal to local translation underdetermines the outcome of bodily collisions, as well as the capacity to apply, and make predictions from, the Cartesian collision rules. Without recourse to a reference frame, one presumably would have to wait until after impact to retrospectively assess, by means of the outcomes of the collision rules, if the local translations amounted to a motion of the body or a motion of the neighborhood—but, if you exploit the (usually) unique post-impact behavior of Cartesian bodies in order to determine which bodies moved, then the local translation story has served no useful purpose.

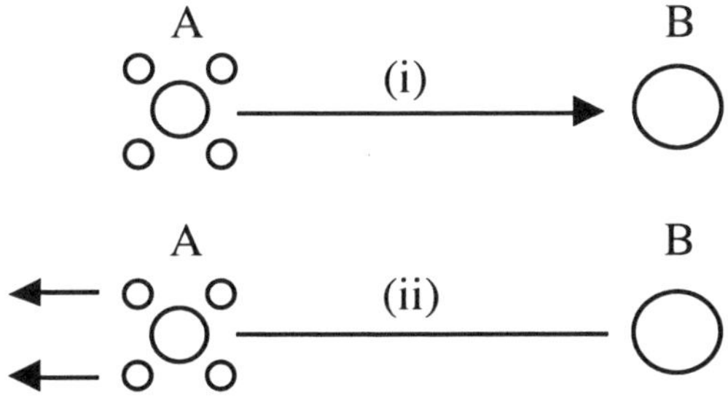

(i) A approaches B
(ii) Distance between A and B is constant

Figure 21. The particles surrounding the stationary body A in (ii) translate in unison to the left.

On the whole, one might hold that Des Chene's "physical trajectory" interpretation of Cartesian motion is immune to these underdetermination problems. In the dilemma posed above, for example, since local translation gives rise to a trajectory, a shrinking spatial distance between the two bodies will eliminate case (i) from consideration. Nevertheless, information on the paths of bodily motion is not sufficient in itself to rescue Cartesian physics from the clutches of underdetermination. To illustrate, consider the example that accompanies the third collision rule: "if [body] B had initially been traveling at six degrees of speed, and [body] C at four degrees of speed, both would subsequently move towards the left at five degrees of speed (Pr II 48)." This sort of physical explanation will continue to create havoc with our various relationalist schemes, but, for the moment, attention should be focused upon Descartes' concept "degree of speed." Overall, it is difficult to see how the impact represented in this example, as in all the others, could be understood by appeal to anything less than a reference frame for measuring the bodies' speeds relative to each other. If speed, as employed in the collision rules, were a measure of local translation alone, then odd situations would arise wherein the approach speeds of bodies differed from their local translation speeds: so, returning to Descartes' example, although body B only approaches at, say, one degree of speed, the "rate of neighborhood change" due to a joint motion of B and its neighborhood results in a local translation speed of six, matching the example—and what would happen in such situations? If the outcome of the collision were dictated by the local translation speeds, than the respective neighborhoods of our two bodies would need to instantaneously harmonize their divergent motions to guarantee the outcome as mandated by rule three; and such long-range, let alone mysterious, cooperation would be asking a lot of the Cartesian plenum, even by Descartes' standards. Finally, and more importantly, the context and elaboration of the collision rules clearly supports the common notion of approach speed, and not rate of local translation, as is evident throughout Descartes' exegesis: e.g., the fourth rule, where B could never move C "no matter how great the speed at which B might approach C (Pr II 49)."

6.2.3. Relationalist Alternatives to (R1).

Despite the best efforts of Garber and Des Chene, therefore, Descartes' analysis of motion remains largely incompatible with his laws of nature—and this is the major weakness of their analysis. Garber concedes this basic point, observing that "without a common framework in which to conceive of the relative motions of more than one body, it is difficult to see how we could give an adequate treatment of the phenomenon of impact (171)." Yet, if this difficulty is acknowledged, then Garber's aforementioned claim that impact rules four and five are not inconsistent (241) is technically correct, but rather misleading. Basically, our commentators seem to be providing the following argument: because it is a fact that a body is, or is not, translating with respect to its neighborhood, there is a real distinction between rules four and five, thus resolving the difficulty. This is only half of what is needed, unfortunately. Although everyone will agree when a body is at rest, or not translating, not everyone will agree on which body moves once a translation occurs. In short, despite the reality of Cartesian motion, a relative motion is ambiguous as regards the individual components of motion, and the collision rules *require* such individual determinations for their correct *application.* As mentioned in section 6.2.1, Leibnizian spacetime is the relational theory best suited to Descartes' actual handling of motion, since the collision rules employ a strong notion of speed (as in the example from rule three). Yet, Leibnizian spacetime can only countenance determinations of "relative speed", and not, without installing reference frames, individual components of speed. The following dilemma is, accordingly, imposed on our Cartesian spacetime: if reference frames are employed to measure the individual speeds of bodies, then the fact that our spacetime is relational means that any reference frame is admissible, which would result in numerous conflicting estimations of bodily speeds, and hence applications of the collision rules; if, however, a privileged class of frames is designated, and the impact rules which conserve "quantity of motion" are only valid relative to them, the spacetime would then appear to be endowed with a structure over and above what relationalism permits.

No matter how unsavory the consequences may be, this last option is apparently the only alternative for the union of a committed (R1) relationalist and the Cartesian natural laws, for it is difficult to perceive how Descartes' physics could be rendered consistent without a set of select view points. Invoking such frames would seem to mark a departure from a strict (R1) reading, nevertheless. In essence, the relationalist could make the following argument: despite retaining Leibnizian structure, the conjunction of this spacetime with the Cartesian natural laws privileges a class of material-based reference frames from which the Cartesian conservation law is upheld. This modified formulation of (R1), which we will denominate, (R1*), thus retains the prohibition on "absolute" quantities of motion (relative to absolute spacetime structure) but allows, via the privileged frames and the Cartesian conservation law, the determination of individual body speeds from the invariant relational quantity "relative speed". Specifically, we can define (R1*) as:

R1*: All motion is the relative motion of bodies, and consequently, spacetime does not have, and cannot have, structures that support "absolute" quantities of motion. But, a privileged class of reference frames may be adapted to the invariant quantities of motion of the spacetime as long as it does not entail a fixed, absolute spacetime structure.

Overall, this sort of "bootstrapping" technique would allow Descartes' "quantity of motion" to pick out the special material-based frames that actually conserve this quantity through measurement, and thus secure agreement on the outcomes of the collision rules.

There remains a certain suspicion, however, that such a strategy merely masks an underlying inertial structure in a relationally palatable fashion. That is, although the spacetime is nominally committed to the non-inertial Leibnizian structure (which does not possess devices for determining the straight-line continuation of paths, like the "covariant derivative" of Newtonian spacetimes, often labeled $\nabla_{\mathbf{V}}\mathbf{V}$, or, $d^2x^\alpha/d\lambda^2 + \Gamma^\alpha_{\mu\nu}(dx^\mu/d\lambda)(dx^\nu/d\lambda)$, in the coordinate frame—see chapter 2), the coupling of this relational spacetime with the Cartesian natural laws essentially constitutes, or presupposes, inertial structure—and inertial structure violates (R1), and by implication (R1*). If this line of argument sounds convincing, then the Cartesian will most likely want to fall back upon (R2), the relationalist thesis which merely denies spacetime substantivalism (i.e., the view that spacetime is a form of substance; Earman, 12). Following Earman's classification scheme:

R2: Spatiotemporal relations among bodies are direct; that is, they are not parasitic on relations to a substantival space that underlies bodies.

As long as space and time are regarded as somehow contingent upon matter (possibly as a sort of "emergent" entity or property of matter?), the Cartesian would thus seem free to invoke as much structure as deemed necessary to explicate Descartes' natural laws, even inertial structure like $\nabla_{\mathbf{V}}\mathbf{V}$. Whether or not (R2) is a consistent relationalist position is the main question, needless to say; and though it is beyond the scope of this essay (see Earman, chap. 6), we will briefly return to this issue in the next section.

6.3. Relationalist Strategies: Reference Frames and Absolute Structure.

Given the results of section 6.2, the Cartesian appears to be led inexorably away from the austere (R1) relationalism to a more moderate variety, whether construed as (R1*) or (R2). One of the most exploited methods of supplying a non-(R1) foundation is to postulate preferred reference frames, either permanently or temporarily fixed, from which the measurements of individual body motions can be conducted. The more fixed these frames become, however, the more they resemble a form of absolute space, as will be discussed. Possibly the first, and most likely the best, of such proposals is Huygens' center-of-mass (-gravity) reference frame.

Huygens, who like most Cartesians was inclined towards an (R1) relationalism (Elzinga, 96), attempted to correct Descartes' collision rules by extending the scope of the first rule, which is the only verifiable hypothesis of the set, to govern the collisions of bodies of all types and sizes. Descartes' first rule asserts that two equally sized bodies moving at identical speed (in opposite directions) will rebound "without having lost any of their speed (Pr II 46)," and thus conserve their quantity of motion. Recognizing what today we would call the Principle of Galilean Relativity, Huygens sought to remedy the deficiencies of Descartes' impact theory by utilizing an identical analysis for the remaining six cases treated in the *Principles,* which would guarantee the conservation of quantity of motion for all interactions. With the discovery of a colliding system's center-of-mass reference frame, Huygens found a means of generalizing Descartes' first collision rule: "if a larger body A strikes a smaller body B, but the velocity of B is to the velocity of A reciprocally as the magnitude [size] A to B, then each will rebound with the same speed with which it came (*Oeuvres Complètes,* vol. 16, 92)." As viewed from the origin of that frame, where bodies preserve their initial speeds after rebound, the ratio of their speeds is reciprocal to the ratio of their sizes. When one body strikes another, irrespective of their size and speed, an observer situated at the center-of-mass perspective will perceive both bodies to recoil in the opposite direction while retaining their initial speeds, and thus quantity of motion. Of the various "reference frame" proposals for mending Descartes' physics, Huygens' project probably comes closest to what we have labeled (R1*), since it is a forthright endeavor to maintain (R1) by using the Cartesian conservation law to isolate a preferred set of plenum-based reference frames; which, in this case, are temporary frames limited to each bodily.

We will postpone the critical examination of Huygens' strategy for reconstructing a Cartesian spacetime until chapter 8. In brief, Huygens utilizes Descartes' first collision rule, which constitutes a member of his third natural law, as the basis of the reconstruction. Yet, many other philosophers have tried a more direct means of grounding Descartes' physics by utilizing and adapting the second laws of nature. One interesting possibility is to rely on the definition of the second law to single out the frames from which straight-line motion recedes. Specifically, Descartes' second natural law states: "all movement is, of itself, along straight lines; and consequently, bodies which are moving in a circle always tend to move away from the center of the circle which they are describing (Pr II 39)." In Pr III 57-58, of the *Principles,* the exact meaning of the passage "move away from the center of the circle" is made clear, for it not along the tangent to the circle that the body strives to recede, as in Newtonian mechanics, but along a radial line directed outward from the center point. Circular motion is the rule and not the exception, moreover, since Descartes believes that all plenum motion ultimately involves a ring of circling bodies, with each body in the ring moving simultaneously into the (internal?) place of the one ahead (Pr II 33; supposedly to avoid the creation of a vacuum). If one correlates these circular motions with the second law, therefore, the tendency towards straight-line motion evidenced by all moving (= circular moving) Cartesian bodies can be referred back, at least in part, to a particular reference frame, i.e., the center point of the circle from which they attempt to flee. For every ring of synchronously circling bodies, there will be a reference frame located at the center point that serves as the perspective for measuring the individual body's "quantity of motion" (which is the measure of this "tendency" or "force" to recede; Pr II 43—see chapter 3 for a more detailed discussion of these issues). Unfortunately, as is often the case with Descartes' physical hypotheses, there is no indication that he remained faithful in the *Principles* to the idea that all bodies must partake in such circular motions; which is especially true of his lengthy discussion of Terrestrial phenomena in Pr IV, where circular motion is hardly mentioned. Likewise, even if this strategy were to supply a system of privileged reference frames for the second law and, presumably, the third law, it would still not remedy the inherent incompatibility of simultaneous plenum motion and a set of collision rules; for, as W. Anderson, has succinctly put it, "the necessary conditions for translation [motion] are contrary to the necessary conditions for collision (Anderson, 220)." If bodies can only participate in synchronous circular motions, when, in fact, are the impact rules ever actualized?

Finally, one can raise the obvious difficulty that a strict (R1) relationalist interpretation of Cartesian motion seems incompatible with the above hypothesis, since a privileged reference frame is picked out among the class of all possible relationally-equivalent frames—namely, the frame that judges the motion to be completely circular is favored since the centrifugal force is coupled to a perfectly circular motion *only* in this frame—whereas other frames would observe a centrifugal force for a motion that is elliptical, non-moving, etc., thus demonstrating that the second law of nature does not take its canonical, or privileged form, in this frame. (In fact, in a frame at rest with respect to the circling body, a force will be manifest on the body in the absence of any motion, as judged from the perspective of that frame.) If one decides to loosen the demand for a strict (R1) relationalism, and accept (R2) or (R1*) instead, then the strategy outlined above may allow for a modified version of Descartes' physics to tolerate a form of relationalism. In section 6.4, and chapters 8 and 9, we will discuss such strategies further.

Other submissions for a Cartesian reference frame have been both less practical and more metaphysically abstract, yet are worth investigating since they exhibit definite (R2) leanings. Representative of one major theme is K. Hübner's proposal, which simply declares God to be the framework from which motions are discerned: *"Descartes' laws of impact describe fundamental occurrences of nature as if seen from the standpoint of God* (Hübner, 130)." This explanation, despite its somewhat incongruous tone, is ostensibly latent in all occasionalist renditions of Descartes' natural philosophy; which holds, roughly, that God is the direct cause of all bodily phenomena (Garber, 299-305). Hübner is led to this conclusion since he believes that the relativity of translation implies that "motion" is merely a mode of thought, and not of extension (128-130). Nevertheless, even if one interprets Descartes' occasionalist-sounding claims, such as "God is the primary cause of motion (Pr II 36)," in a literal fashion, and it is unclear that one should, Hübner's tactics are fundamentally implausible in that they procure an ontological foundation for Descartes' natural laws at the expense of the epistemological. When Descartes listed the factors that complicate the application of his impact rules, such as lack of solidity and plenum interference (Pr II 45), he did not further add that only God could provide the correct calculations. Possibly sensing this dilemma, Hübner's solution is rather Platonic: "the true changes [in bodies] are, namely, those given in divine revelation, and not the merely apparent ones given through the senses or determined in accordance with some arbitrary relativity as a [mode of thought] (130)." Despite the essential implausibility of appealing to "divine revelation" to ground Descartes' physics, we will return to a more promising basic thread in Hübner's argument in section 6.4.

Furthermore, as the reference frames postulated by these commentators become more and more permanent, the Cartesian spacetime grows inevitably more "absolutist". This is nicely demonstrated in the case of R. Dugas, who entertained the notion that the material world prior to the formulation of the three material elements and planetary vortices, i.e., before God imparted a conserved quantity of motion, serves as the sole preferred frame (Dugas, 196; for the Cartesian cosmological creation story, see Pr III 46). Des Chene's notion of a "physical trajectory", explained above, also appears to invoke this richer spacetime structure, particularly in his repeated denials that Cartesian motion obligates "a local frame of reference in relation to which motion is defined (262)." Des Chene repudiates such arbitrary frameworks for measuring motion, arguing instead that Descartes "is trying to explain what is true of a thing at each moment of its *absolute* motion (272)." A view characterized by "the instantaneous, punctual *translatio* from the vicinity of one set of bodies to another, and the thought that by gluing together the spaces successively occupied by a body one can construct a physical trajectory to which the relative conception of place is irrelevant (272)." Ironically, in a later section, Des Chene cautions against thinking of Descartes' theory "as being like Newtonian absolute space (374)"; yet his "trajectory" hypothesis follows, practically verbatim, Newton's reasoning in the *De gravitatione* (section 6.2), which rejects Cartesian motion for lacking this absolute structure! For both Newton and Des Chene, motion requires a spacetime that can identify the same spatial position across time, thus ensuring a coherent groundwork for measurements of speed, trajectory, etc.; and only the Full-Newtonian spacetime has the requisite structure to accomplish this task (being thus equipped with a spacetime "rigging" to identify such positions; see Friedman 1983, 74).[3]

The preceding analysis should not be taken as necessarily denouncing Des Chene's hypothesis, however, for there is much in Descartes' approach to the problem of space and motion which suggests a commitment to absolute space, if only to its structure via (R2) and not to its ontology. In fact, Des Chene's interpretation helps to belie the knee-jerk tendency to categorize historical theories of space and motion as either clearly in the strict (R1) relationalist camp or clearly in the substantivalist camp. If Descartes' espousal of a relationalist theory is viewed more as a repudiation of the idea that space is an entity independent of matter, as certain of his predecessors like Patrizi believed, then it is possible to read Descartes as having endowed space with a more robust structure than has been commonly accepted. As presented in Grant (192), there are historical precedents prior to Descartes for the belief that space (or vacuum), although different from matter, can only exist as long as matter exists; namely, John Philoponus. Also, in common with his Scholastic predecessors, Descartes' theory of motion as change of (external) place rules out the possibility that the entire plenum could, in unison, translate or rotate: since the plenum as a whole has no "contiguous neighborhood," it cannot be said to move. The line of thought leading from Patrizi, through More to Newton, on the other hand, clearly allows for such "global" motions of the material world, for space is a "thing" (not necessarily either substance or property) *independent* of matter. With respect to global material motion, consequently, Descartes is a straightforward relationalist, but Des Chene's position intimates that he may have endorsed an even richer structure for space, possibly as potent as Newtonian spacetime, just as long as it is acknowledged that it cannot exist apart from matter.

Unfortunately, this hypothesis, which Earman has dubbed the "property view of space (14),"[4] does not appear to be corroborated by the evidence from the texts, since Descartes repeatedly claims that "the names 'place' or 'space' do not signify a thing different from the body which is said to be in the place (Pr II 13)." Overall, Descartes' view is quite nominalist, for he seems to regard space as a sort of conceptual abstraction, much in the way that numbers are abstractions: "quantity [extension, volume] does not in fact differ from the extended substance except insofar as our conception of it is concerned; similarly, number does not differ from the thing which is numbered (Pr II 8)." A conceptual abstraction does not amount to a property theory of space, needless to say, nor does it presumably have the capacity to ground the individual motions of bodies. Thus, unless textual evidence is forthcoming that can verify that Descartes equated his conceptual abstractions with some form of higher-level structural feature of matter (that is capable of grounding motion), this theory must remain merely an intriguing suggestion.

6.4. Dynamical Considerations.

In the previous two sections, interpretations of Cartesian motion that fit the (R1) category, as well as some non-(R1) formulations, have been shown to be unsatisfactory. As in the cases of both Garber and Des Chene, the failure to take into account the demands of Cartesian dynamics (i.e., natural laws) is the source of most of the difficulties (although some interesting hypotheses are simply inconsistent with Descartes' texts). This section, on the other hand, will present a resolution to the problem of Cartesian motion which draws its support from Descartes' dynamics, especially from a current debate centered upon its ontological status of "force." In common with the natural laws, Cartesian force has generally not been considered a means of resolving the troubling issues surrounding the relativity of motion. Rather, force has traditionally been discussed by commentators after having grappled with the pure "kinematic" concepts of motion; i.e., translation, reciprocity, neighborhood, etc.. As in both Garber's and Des Chene's texts, the examination of force is largely separate from the kinematical exegesis. There are many reasons for this general approach, most notably Descartes' general avoidance of the relativity issue in his scientific texts. Nevertheless, there are good reasons for reversing the course of this "normal" interpretation of Cartesian physics, and thus viewing Descartes' dynamics as primary or essential to a proper understanding of his kinematics.

Before proceeding further, however, we need to reexamine the fourth and fifth impact rules to discuss the apparent distinction between a force of resistance and force of motion (see, also, chapter 3). In rule five, the larger body B transfer as much speed to the smaller stationary C in order to allow them both to move in B's direction at the same speed. This suggests that B's force for "proceeding", or motion, is simply the product of its size and speed, as one would expect given the conserved quantity of motion. But, Descartes also postulates a force of "resistance": "a body which is at rest puts up more resistance to high speed than to low speed; and this resistance increases in proportion to the speed (Pr II 49)." This force becomes manifest in the fourth rule, where the larger resting C deflects the smaller moving B back along its path. Since C is at rest, its force of resistance is presumably measured by the size of C and the speed of B, for only then is its force greater than the moving B's force of motion.

With respect to bodily force, moreover, commentators are divided as to its ontological status: some, such as M. Gueroult and A. Gabbey, conceive force, in at least one important sense, as being *in* the objects themselves as a feature or mode of body; while others, like Garber and Des Chene, relegate it to a sort of derived phenomenal effect of God's sustenance of bodily speed and size, and thus not present within extension. Regardless of which side of the dispute one favors, a potential resolution of the kinematical puzzle of relative motion is contained in this deliberation on the dynamic status of quantity of motion.

If one sides with the Gueroult/Gabbey interpretation, for instance, then Descartes' handling of the conserved quantity of motion in the collision rules, which seems to posit the separate forces of rest and motion, demonstrates that these forces "are immanent in 'nature' or extension . . . and can be calculated at each instant for each body, according to the formula *mv* (Gueroult, 198)." Likewise, Gabbey reckons "at the practical level of physical investigation, forces—whether of motion or rest—are real causes in their own right and distinct from motion and rest (Gabbey, 238)." Interestingly, by placing the force in the body, whether as a mode or some aspect of extension, this maneuver effectively locates the long sought after reference frames for determining the individual motions of impacting bodies—the frame is an intrinsic feature of bodily forces, principally rest. Since a resting body puts up a force of resistance comparable to its size and the *speed of the approaching body,* its internal force constitutes a built-in method of determining not only its state of motion, but the speed of the second impacting body. Thus, a resting body's force acts as a reference frame for overcoming the relativity of Descartes' reciprocity of transfer hypothesis: regardless of which bodies are deemed to move once a translation occurs, and given the existence of non-translating bodies in the Cartesian plenum (which a safe assumption), these resting bodies have an intrinsic capacity to determine the individual speeds of the bodies they meet on impact, and this capacity constitutes a platform that, at least hypothetically, can be extended as a reference frame to determine the individual states of motion of all plenum bodies. In this context, a "resting" body is to be understood according to the concept of "proper-rest" described above: i.e., as *non-translation* with respect to the *contiguous neighboring bodies,* which is an invariant (non-relative) feature of the spacetime. This eliminates, of course, the problem that "vulgar rest" (a.k.a. Galilean relativity) would pose for this thesis. Since, if manifestations of the rest force were relative to any and all inertial perspectives, then a host of conflicting determinations of the force would be ascribed to the same body. Once again, the non-(R1) nature of Descartes' theory is in evidence. Nevertheless, by utilizing this "rest force" process, one can secure a foundation for the measurement of the collisions rules, and hence natural laws.

Moreover, even if one rejects these bodily forces as actual properties of bodies, and adopts the Garber interpretation instead, it is still the case that resting bodies behave as if they possess forces of rest, and hence they will retain the reference frame ability detailed above. More carefully, Garber (283) and Des Chene (327) both contend that God sustains a certain quantity of motion in the world according to the plan of Descartes' natural laws, and thus there is nothing in bodies except their extension. Yet, God's maintenance of the Cartesian laws ensures that resting bodies (which are not problematic on the local translation scheme of section 6.2) will have the capacity to measure the approaching speeds of impacting bodies, and this is all that is needed to resolve the problem of relative motion. Since God is doing the work, one might be tempted to claim, along with Hübner, that God is now the preferred reference frame; and while this may be superficially correct, it is also true that, under the ideal conditions stipulated in the *Principles,* Descartes assumes that the individual motions of bodies are determinable using normal empirical means, without recourse to divine revelation.

This last ingredient—the ability to gauge the speed of the approaching body, as well as the body's own state of motion—is what has been needed all along to secure a foundation for the collision rules free of the relativity problem, and it looks as if Descartes has assumed it as a basic constituent of Cartesian bodies all along. Consequently, Garber and Des Chene are correct in denying that Descartes favored a strong (R1) relationalist theory; but, at least on this estimation, they have reached the right conclusion while simultaneously neglecting the source of the main difficulty, namely the need to supply the privileged reference frames to secure a consistent relational theory of both kinematics and dynamics. (In all fairness, Des Chene's absolute trajectory hypothesis did try to meet this challenge.) Given the discussion in section 6.2, Garber's and Des Chene's accurate interpretation of Descartes' "reciprocity of transfer" hypothesis is insufficient to resolve all the difficulties associated with relational motion, particularly as regards the collision rules (and they both seem to admit this point in numerous asides throughout their texts). Oddly enough, the indispensable mechanism for determining the relative speeds of bodies, as mandated by the impact rules, appears to be embodied within (resting) Cartesian bodies as a result of

Descartes' own handling of his conservation law. Put differently, one way of summarizing the pitfalls of most, if not all, earlier examinations of Cartesian motion is that commentators have (quite naturally) tried to resolve the kinematic problem of the reciprocity of transfer before addressing the inherent dynamic weaknesses of Descartes' physics; whereas a more adequate solution may lie in attending to the dynamical aspects prior to, or in conjunction with, the kinematical details of Descartes' theory.

One major problem with this explanation, unfortunately, is that it ostensibly undermines the purpose and importance of the reciprocity of transfer thesis. If the conservation law can ground the individual motions of bodies on its own, then there is no apparent reason for Descartes' local translation story; unless, of course, it is simply to caution against regarding motion as something over and above extension, i.e., a dynamic property *in* bodies, which is a reading of its intended meaning that would seem to favor the Garber/Des Chene interpretation above. In other words, motion and rest, considered apart from a conserved quantity of motion, are merely relations between a body and its contiguous neighborhood which are both symmetric and kinematic. Once God introduces the (dynamic) element of a conserved quantity of motion, however, the symmetry of the kinematic relations is broken, and a preferred manner of viewing bodily motions is imposed on the spacetime observer.

Moreover, if we concede the consistency of (R1*), then the reference system described above would appear to vindicate this form of relational structure, since it is the conserved quantity of motion that guarantees the existence of the reference frames for determining the individual states of bodily motion. The conserved quantity of motion, as a whole, "bootstraps" the privileged reference frames required to give a coherent relational measure of that quantity, as well as the specific outcomes of the seven collision rules. Consequently, it is not necessary to view the reference frame process described in this section as mandating a special bodily rest force. In fact, an alternative rendition of our reference frame technique could forego the need for the resting force to instate the frames, altogether; and, instead, simply declare that the conjunction of the Cartesian conservation law and the "relative speeds" among all impacting bodies always selects a unique perspective for determining their individual motions. In short, rather than relying on some nearby resting body to measure motion, and thus seemingly invoke a force inside bodies, the conserved force could directly individuate bodily motions relative to unique reference frames which may, or may not, be attached to one of the colliding bodies. This version of our reference frame strategy is hence closely akin to Huygens' center-of-mass frame method of upholding the Cartesian conservation law and collision rules. Yet, whereas Huygens utilized the specific outcome of the first collision rule as the basis of his reconstruction of Cartesian dynamics (see chapter 8), the story outlined above only uses the conserved quantity of motion to directly select the chosen perspectives to maintain the quantity: it does not attempt to rework the precise predictions of the impact rules, as was one of Huygens' main intentions. On the other hand, the reference frame technique developed in this chapter is quite similar to Huygens' method in that the instantiation of the preferred frames occurs within the context of individual bodily collisions of a finite duration. Since each impact only spans a finite temporal period, the frames that conserve the quantity of motion for the colliding bodies would not be permanent features of the spacetime (i.e., exist at all past and future moments of time), but only temporarily fixed reference frames. In chapter 8, we will discuss the nature of temporarily fixed reference frames at greater length.

As for textual evidence supporting our dynamics-based "rest force" maneuver, the very details of the collision rules straightforwardly confirm that resting Cartesian bodies have this capacity (i.e., to untangle the relational difficulties by measuring approach speeds, as above). Likewise, the dynamically-conceived “tendency” of bodies to flee the center of their circular motion, which thus establishes an (R2) or (R1*) reference frame at the center point (see section 6.3), greatly assists in establishing the central importance of dynamical considerations in Descartes’ physics of motion. Therefore, the impact rules and the presentation of circular tendencies in Pr II both provide strong indirect evidence for the general claim that Cartesian dynamics can resolve the kinematic problem. But, a more direct instance of dynamical consideration taking precedence over the kinematical would greatly strengthen the arguments of this section. While discussions of this sort are extremely rare in the Cartesian corpus, the correspondence between Descartes and Henry More, first introduced in chapter 5, forms a supremely important piece of confirming evidence for the proposal outlined above. Recalling the context of the discussion, a challenge by Henry More prompted Descartes to defend his relationalist views by means of a remarkable hypothetical case (which, once again, has been largely neglected in the literature). If transfer of a body is merely reciprocal, More asked how one should regard the case of wind blowing through a tower window (5 March 1649, AT V 312): Is this motion reciprocal as well, so that the air can be viewed at rest and the tower in motion? Rather than answer More's question as kinematically presented, Descartes tries to resolve the problem by invoking a different, dynamics-based example. This example involves the force or "strength" (*vires*) of two men attempting to free a grounded boat; with one on board pushing against the shore, and the other on shore pushing against the boat:

> If the force (*vires*) of the men is identical, the effort of the man on the shore, who is thus connected to the land, contributes no less to the boat's motion that the effort of the man on the boat, who is transported along with it. Therefore it is obvious that the action by which thc boat rcccdcs from the shore is equally in the shore as in the boat (15 April 1649, AT V 346).

In essence, Descartes' "boat" example, as we will call it, applies a *dynamical* solution to an original *kinematical* problem: the reciprocity of the transfer of the boat and shore, which is a rather troubling kinematical consequence of his espoused relationalism, is reinterpreted as a sort of "reciprocity of force" between two possible dynamic sources of the motion, namely the push of each man.

One might be tempted to agree with Shea (1991, 323) that Descartes' explication of the "boat" example unwittingly conflates the kinematic and dynamic elements of his theory. This may be true, but it does not undermine the importance of the above passage for our present purposes. As *literally* interpreted (which is always a virtuous appeal by a commentator), Cartesian dynamics would appear, at the very least, intertwined with Cartesian kinematics, if not (in this case) taking precedence over the latter (see chapter 5). The reciprocity of transfer "process", broadly construed, now appears to encompass force as much as it involves kinematics. In fact, Descartes' reply to More seems to tacitly acknowledge that it is "really" the boat that is in motion, and not the shore, which would seem to be the motivation underlying the switch to a different example: i.e., it is the reciprocal nature of the force (of pushing) that causes or generates the individual motion of the boat, which is a relationally "legitimate" state-of-affairs not present in More's original case (or so Descartes believes). Of course, Descartes is not implementing a rest force to break the symmetry of the kinematical transfer (as argued above)—but he is appealing to the dynamics of the situation *in general* to resolve this relational difficulty. Thus, in the letter to More, Descartes' resolution of the kinematical "reciprocity of transfer" problem *falls back upon the dynamics* of his theory of motion in the exact same way that his kinematical collision rules employ an intrinsic rest force (to measure approach speeds) or the second natural law picks out a privileged reference frame by means of a dynamical "tendency" to flee a privileged circular motion in the spacetime. In the "boat" example, however, the primary role of dynamics (in solving kinematical puzzles) is much more directly presented and evident (if possibly more confused) than in its analogous use in the collision rules. The importance of Descartes' "boat" example should not be underestimated, moreover: not only does it have a chronological advantage over earlier work (being one of his last discussions of natural philosophy; spring, 1649), but it is one of the only known instances where he was actually pressed for the details of his brand of relationalism (by way of a counter-argument).[5]

Finally, the overall results of this section should be placed within the historical context of another relationalist strategy not yet surveyed in our investigation. By holding motion to be generally relational, but that individual determinations of motion are an intrinsic fact of bodies, this interpretation of Descartes' places him close to Leibniz's view of space and time, or at least to one of Leibniz' many themes on these issues. In the "Discourse on Metaphysics," he states:

> For considering only what it means narrowly and formally, that is, a change of place, motion is not something entirely real; when a number of bodies change their position with respect to each other, it is impossible, merely from a consideration of these changes, to determine to which of the bodies motion ought to be ascribed. . . . But the force or the immediate cause of these changes is something more real, and there is a sufficient basis for ascribing it to one body rather than another. This, therefore, is also the way to learn to which body the motion preferably belongs (Leibniz 1970, 315).

Earman concludes (132) that this form of reasoning falls within the (R2) class, for it requires a structure for spacetime richer than (R1) relationalism allows (but not (R1*), since, if we grant its consistency, Leibnizian spacetime structure would be preserved, as above). The theory of Cartesian motion developed in this section closely follows the Leibnizian argument, as well. Like Leibniz, Descartes' theory would entail that motion, kinematically conceived, is relational (as in the reciprocity of transfer), but that the dynamical aspects of bodies can assist in breaking the relational symmetry in order to reveal which ones are truly in motion. Whether regarded as *in* bodies, according to Leibniz or Gueroult/Gabbey, or deemed to be the phenomenal product of God's sustaining act, according to Garber/Des Chene, bodies have a capacity *born of conserved forces* (i.e., the rest force, unique circular striving) to reveal individual states of motion, contra (R1). As an interesting aside, L. Sklar's theory of "absolute acceleration", as an internal feature of bodies without need of substantival space, would also seem to correspond to this view (230). But this analogy should not be pushed too far, since, unlike Leibniz and Sklar, force is not required to be an intrinsic property of Cartesian bodies. Notwithstanding these differences, with this last set of references it has been thus surreptitiously demonstrated that nearly every modern hypothesis of space and time, from strict relationalism to absolute spacetime structure, has an analogous interpretive counterpart in the philosophy of Cartesian space and motion.

6.5. Conclusions.

In retrospect, it is fair to say that Cartesian physics has been largely neglected, if not outright ignored, by twentieth century philosophers of space and time. While there have been no shortage of attempts to encapsulate Newton's natural philosophy in the modern mathematical formalism (starting with the pioneering work of Cartan), Descartes' theories have not been as fortunate. One of the goals of our overall project has been to partially reverse this trend by demonstrating the added insights that the modern formalism can provide the Cartesian researcher. In particular, the preceding analysis has strived to clarify what most commentators have often incompletely or vaguely perceived; i.e., that a strict (R1) relationalism is incompatible with the Cartesian natural laws. Nonetheless, most commentators have been even less successful in perceiving that there are "weaker" formulations of relationalism that may adequately ground Descartes' physics. This realization has ultimately prompted the search, as disclosed more fully in this chapter, for the elusive privileged reference frames that can provide a sound basis for a relational (R1*) or (R2) spacetime. Specifically, most commentators have also overlooked the fundamental importance of Cartesian dynamics in resolving the kinematic obstacles of the "reciprocity of transfer" hypothesis—a startling feature of Descartes' theory directly supported in his letter to More. Moreover, as argued, Descartes' handling of the collision rules strongly suggests that Cartesian bodies have a capacity, via the conservation law, for determining the individual motions of both themselves and their co-impacting partners, thus establishing the groundwork for the privileged reference frames as mandated by (R1*) or (R2) relationalism. In the remainder of Part III, we will explore a number of different formulations of Cartesian spacetime that also attempt to reconcile Descartes' natural laws with his theory of relational motion. These versions of Descartes' theory will not necessarily employ the specific dynamics-based method of upholding Descartes' natural laws as constructed in this chapter (although Huygens' hypothesis is closely akin to this approach, see chapter 8, as well as the kinematics of mechanics procedure in chapter 9).

ENDNOTES

[1] J. Bennett, "Space and Subtle Matter in Descartes", in *New Essays on the Rationalists,* R. J. Gennaro and C. Huenemann, eds. (Oxford: Oxford university Press, 1999), 3-26.

[2] By defining motion as change of "neighborhood", Descartes could thus claim that the Earth remained at rest in its band of vortex particles (since the contiguous particles did not change), as required by Church doctrine, yet still claim that this band *as a whole* moved around the sun. See, Pr III 28-29. I agree with Garber (1992, 186-188) that Descartes' theory of motion in the *Principles* is his genuine view and cannot be simply reduced to a political expedient.

[3] Des Chene somewhat confuses the issue by stating that "the distance between two points (or spacetime points) is likewise an invariant (270, fn. 13)," after discussing the (supposedly?) metric invariants of spacetimes that license Galilean (Neo-Newtonian) and Special Relativistic kinematics. But, spatial distance is not an invariant between non-simultaneous (event) locations in Neo-Newtonian and Special Relativistic spacetimes; in fact, it is not even a meaningful concept in such spacetimes. Only Full-Newtonian spacetime has a well-defined notion of spatial separation between non-simultaneous event locations (Sklar 1974, 207), which Des Chene appears to be assuming in his physical trajectory concept.

[4] Technically, the property view of space violates both absolutism and (R2), since (R2) also contends that bodies cannot have individual properties of spatiotemporal location. Yet, in rejecting substantivalism and the independence of space from matter, it is much closer in spirit to relationalism, and will thus be included within the (R2) category.

[5] The "boat" example also raises interesting questions about the epistemology of force (which are beyond the scope of this book, however). Although the conservation of force in the totality of matter is grounded in God's "immutability", and thus has a claim to "a priori" status (Pr II 36-37), Descartes' example apparently suggests that the forces involved in particular motions (the boat) can be identified and delimited, to some degree, through observation (the reciprocal push of the men). See, Normore (1993) and

Nadler (1990), for some recent work on the overall epistemological status of the conservation principle and natural laws.

CHAPTER 7

THE KINEMATIC LOGIC OF RELATIONAL TRANSFER: AN UNWRITTEN CHAPTER IN THE HISTORY OF CARTESIAN MOTION

While nearly every aspect of Descartes' much-maligned theory of motion has been carefully scrutinized by commentators, there are a few hypotheses advanced in the *Principles of Philosophy* that have been, oddly enough, largely overlooked. Essentially, these criteria constitute an amendment to the Cartesian doctrine of relational motion: i.e., that motion is the reciprocal translation of a body and its contiguous neighborhood (of surrounding plenum bodies). In his neglected refinements to this view, Descartes argues that there is a way to determine which of the two reciprocally translating bodies is *really* in motion. When two smaller bodies move in opposite directions on the surface of a larger body, he explains, the kinematics of the scenario rule out the possibility that it is the larger body that is in motion. Moreover, a body is in motion if its whole surface, and not merely a portion of its surface, moves relative to its neighborhood. The importance of these "supplementary" criteria cannot be overestimated, for they would seem to compromise the "strong" form of relational motion normally attributed to Descartes; since, to be more specific, these criteria would appear to challenge Descartes' principal judgment that motion is a purely reciprocal change of a body's contiguous neighborhood.

In this chapter, we will explore the Cartesian theory of motion in the light of the crucial addendum that appear in the *Principles*. Although Descartes' arguments are deficient in various ways, as will be demonstrated, the few attempts by commentators to deal with these hypotheses—mainly, D. Garber and M. Gueroult—have not correctly diagnosed their underlying weakness, Overall, despite their ingenuity, it will be demonstrated that Descartes' criteria cannot rescue his brand of natural laws from the inherent limitations of his strong relational account of motion. Consequently, while not without merit, Descartes' supplementary hypotheses cannot serve as the basis of a consistent Cartesian spacetime (which is the goal of Part III). namely, their inability to consistently ground the Cartesian natural laws. While section 1 will offer a brief description of Descartes' overall theory, sections 2 through 4 will evaluate the strengths and weaknesses of Cartesian motion provided his important additional stipulations.

7.1. Introduction

As explained at length in chapters 2 and 6, Descartes' *Principles of Philosophy* contains an analysis of motion that exhibits strong relationalist tendencies. In this work, which presents his most extensive discussion of this phenomena, Descartes defines motion (*motus*) as "*the transfer [translatio] of one piece of matter or of one body, from the neighborhood [vicinia] of those bodies immediately contiguous to it and considered at rest, into the neighborhood of others* (Pr II 25)." Since Descartes attempts to distinguish his "proper" conception of motion, as change of neighborhood, from the "vulgar" conception of motion as a change of "place" (*locus*); he holds that "(internal) place" be defined as the situation of the body relative to some arbitrary set of, usually resting, distant bodies. Thus, any attempt to stipulate that the surrounding bodies are "at rest" must remain purely arbitrary, since "we cannot conceive of the body AB being transported from the vicinity of the body CD without also understanding that the body CD is transported from the vicinity of the body AB (Pr II 10-15, 24-29)."

Recalling the discussion from chapter 3, the non-relational factors at work in Cartesian motion are nicely exhibited in Descartes' handling of his collision rules, which constitute the specific instances of his third natural law. The inconsistencies are most conspicuous in the case of rules four and five: briefly, in the fourth rule, a large object remains at rest during impact with a smaller moving body, and simply deflects the smaller body back along its path (Pr II 49); whereas in the fifth rule, a large body will move a smaller stationary object, "transferring to [the smaller body] as much of its motion as would permit the two to travel subsequently at the same speed (Pr II 50)." From a relational standpoint, however, rules four and five constitute an identical collision, since they both involve the interaction of a small and large body with the same relative motion (here, a simple speed difference). Therefore, because they represent an identical scenario, the consistency of Descartes' alleged relationalism is placed in serious doubt.

D. Garber has challenged this conclusion, nevertheless, by pointing out a non-relational facet of translation that can greatly assist the Cartesian in resolving the dilemma of impact rules four and five:

> For Descartes, the case in which [a body] is in motion is physically distinct from the case in which it is at rest. And so, for him, the situations described in [rule four] and [rule five] are not mere redescriptions of one another; one cannot arbitrarily designate which of two bodies in relative motion is in motion and which is at rest (241).

In other words, since there is a means of discerning rest from motion—translation of neighborhood or no translation of neighborhood—the fourth and fifth rules can thus be individuated without transgressing relationalist tenets (Des Chene 1996, makes the same point, 297-298). Or, to put it differently, while translation *per se* is relational, such that a translation does not provide a means of determining, say, which of two bodies moved, it is nonetheless a fact that a body is *really* at rest, and not in motion, if *no* translation takes place—and this fact can be put to advantage by the Cartesian. We will return Garber's important observation in the next section, but first we need to examine the details of the additional hypotheses that Descartes amended to his theory of motion.

7.2. The "Single Body" and "Whole Body" Displacement Hypotheses.

The first of the refinements to his reciprocity of transfer thesis appears in Pr II, Article 30, of the *Principles,* which addresses the reason "why the motion that separates two contiguous bodies is attributed to one rather than to the other." Descartes provides a lengthy example to demonstrate his point:

> We do not think that a body is moving unless it moves as a whole, and therefore we cannot understand that the whole earth moves simply for the reason that some of its parts are transported from the vicinity of other smaller bodies with which they touch; and since we frequently notice many similar transferences that are contrary to one another. For instance, if the body EFGH is the earth, and if, on its surface, body AB is transported form E towards F at the identical time that body CD is transported form H towards G; then although we know that the portions of the earth contiguous to body AB are transported from B towards A, and the action of the transference cannot be either different or weaker in the parts of the earth as in the body AB; we do not by that understand that the earth moves from B towards A, or from the East towards the West; since by the fact that those parts of it which touch the body CD are simultaneously being transported from C towards D, we would also have to understand that the earth moves in the opposite direction, that is, from West to East; and these two assertions are contradictory. Therefore, unless we deviate too much from the traditional manner of speaking, we should say that the bodies AB and CD, and other similar bodies, move; and not the earth (Pr II 30).

Put briefly, the relationally symmetric translation between a body and its contiguous neighborhood is broken by the motion of two smaller bodies, in opposite directions, along the surface of a larger body. Since it is impossible for the larger body to be moving both east to west, to account for the motion of AB, and simultaneously from west to east, to explicate the motion of CD, it must be the case that the larger body is at rest. That is, if the larger body is moving east to west relative to AB, then it must be moving east to west relative to CD as well, since both AB and CD are located on the surface of EFGH at the same time (i.e., the time of the motion in question). If the portion of EFGH contiguous to AB were to move east to west relative to AB, and the portion of EFGH contiguous to CD were to simultaneously move west to east relative to CD, then the resulting oppositely-directed motions must shear EFGH in half. Thus, assuming EFGH remains intact, the two smaller bodies must move in opposite directions upon the surface of the resting larger body (see Figure 22).

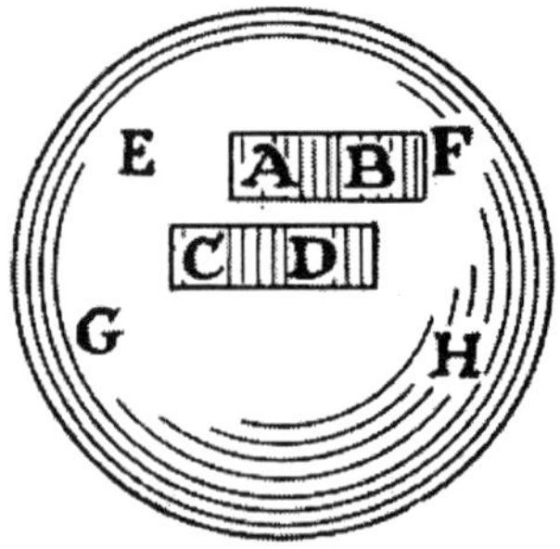

Figure 22. A reconstruction of Descartes' illustration for the example utilized in Pr II, Article 30, of the *Principles*.

This criterion, which M. Gueroult (1980, 208) has dubbed "single body" displacement, constitutes an important advance in the development of Cartesian kinematics (or so it would seem at first glance). As described above, Descartes' physics requires some means of determining the individual motions of bodies if it is to remain consistent with his natural laws. Whether or not a body is at rest or in motion is a non-relational feature of Cartesian spacetime, as Garber correctly notes: i.e., whether or not there is a translation, or no translation, between a body and its contiguous neighborhood is an "invariant" fact for all observers. Nevertheless, the sheer fact that a body is in motion, or lacks motion, is not sufficient in itself to underwrite the Cartesian natural laws, let alone the collision rules. To ground the collision rules effectively, such as rules four and five, one needs to determine if the two bodies are approaching one another; and this property of the impact phenomenon is not captured by sole reference to the neighborhood of each individual body. Even provided relational motion between a body and its neighborhood of contiguous bodies, it is still not possible to determine which collision rule the impact will fall under, or if the bodies will even collide. Suppose, for example, that the two bodies are separated by a fixed spatial distance, and that one of the bodies is, and the other is not, undergoing a reciprocal translation with respect to its contiguous neighborhood. Given this scenario, it is not possible to determine if; (i) the "translating" body approaches the "non-translating" body, or (ii) the spatial distance between them remains fixed and the translating body simply experiences a change of neighborhood (i.e., the neighborhood moves but the body does not—see chapter 6, and Figure 21, for a discussion of this same problem in a slightly different context). Therefore, an appeal to the reciprocity of transfer thesis results in an underdetermination of the outcome of bodily collisions, as well as the capacity to apply, and make predictions from, the Cartesian collision rules.

Nevertheless, it is at this point that Descartes' single body displacement hypothesis can prove itself to be quite useful. If it is possible, following this hypothesis, to determine if a body is in motion by analyzing the motions of the smaller particles on its surface, then the underdetermination dilemma imposed by scenarios (i) and (ii) can be successfully resolved: in short, option (ii) will be eliminated because the translation of the body and its contiguous neighborhood will reveal upon closer inspection to be solely due to the motion of the neighborhood, rather than the body. That is, the larger body will manifest (possibly numerous) pairs of smaller bodies moving in opposite directions along its surface (as in Figure 22); and, with the discovery that both bodies in case (ii) are actually at rest, this scenario does not thus amount to an impending instance of collision rule four or five. There are, of course, many difficulties associated with this theory.[1] But, before assessing the success of the single body, or SB, displacement hypothesis, it will be useful to examine the second criterion that Descartes offered in striving to overcome the limitations of his theory of motion.

One of the first commentator to recognize the importance of SB displacement was M. Gueroult (1980), who also directed attention to the second Cartesian hypothesis, which he deemed, "whole body" (WB) displacement. In this addendum to his theory of motion, Descartes argues that if a translation occurs between two bodies, but only one of the bodies undergoes a change of neighborhood over its *entire* surface, then the motion must be attributed to this body (rather than to the body that only experienced a partial change of neighborhood). Descartes introduced this criterion in large part to counter the Tychonic view that placed a stationary earth at the center of the rotating planetary spheres. He reasons that only the inner surface of the planetary sphere, which is contiguous with the earth's surface, and not the sphere's outer surface (which is contiguous with the next planetary sphere in the series), experiences a change of neighborhood with respect to a translation between that inner surface and the earth. Motion should be assigned to the earth, and not the planetary sphere, since the whole surface of the earth undergoes a change of contiguous neighborhood (whereas only the inner surface of the earth-enclosing sphere changes its neighborhood in this motion). As a result, motion "must be solely attributed to the [planetary spheres] and not the planets; exactly as the partial transfer of water and air that occur on the earth's surface are normally attributed, not to the earth, but to those portions of the air and water that are transported (Pr III 28)."

Unfortunately, this new criterion would not appear to be as successful as its SB counterpart in resolving our previous problem of relational motion, for it cannot dispel the underdetermination dilemma examined above; i.e., the problem of which scenario, (i) or (ii), is the actual outcome of a translating and non-translating body (as originally specified in collision rules four and five). The WB displacement criterion cannot resolve the dilemma because the majority of the smaller particles will only experience a change of contiguous neighborhood on the side facing the larger body; and hence these smaller particles will be deemed "at rest". Put differently, since many of the smaller particles will flow together in a joint, synchronous motion (en masse)—that is, the particles will be at rest relative to one another (but not relative to the larger body)—only the sides of the particles contiguous with the larger body will undergo a "change of place." The larger body will be categorized as "in motion", on the other hand, due to the complete change of its contiguous neighborhood (i.e., all the minute particles that surround it are in motion relative to its surface). Unlike the SB hypothesis, consequently, the WB criterion will fail to identify (eliminate) instances of scenario (i) from scenario (ii). Since both cases (i) and (ii) depict a translating and non-translating body, these scenarios will be generally indistinguishable according to the WB displacement hypothesis, and thus it will be impossible to identify which one of these hypothetical scenarios is operative unless some outside frame of reference is introduced (or one simply waits until the bodies do, or do not, collide).

Of course, once an outside perspective is introduced to distinguish these impact scenarios, such as a reference frame located on one of the colliding bodies, the *conjunction* of this frame and the WB or SB hypotheses essentially equips the spacetime with a privileged reference frame, in direct violation of the tenets of Descartes' strict relationalism, labeled (R1) in chapter 6. A strict (R1) relationalism is undermined since the supplementary criteria will uniquely determine whether the particular frame is either *really* at rest or *really* in motion (since it is attached to a body that, via the criteria, is either *really* at rest or in motion), and thus the frame can now serve as a means of procuring the same determinations of motion for all Cartesian bodies. For many a latter-day Cartesian, however, this form of "weak" (or "reference frame") relationalism, either of the (R1*) or (R2) variety, may be a consistent and acceptable alternative to Descartes' more austere variety (see chapter 6). On this weakened form of relationalism, motion is still a reciprocal relation among bodies, but there are means *other than absolute or substantival space* for determining their individual states of motion. Yet, it is an open question whether or not such a relational variant is compatible with the spirit of Descartes' theory, especially when one recalls that the supplementary criteria, which were intended to shore up the difficulties in his original account of motion, askew all recourse to outside frames of reference for determining such individual bodily states. Descartes attempts to rely on a form of "kinematic logic" in these criteria, as will be discussed below, rather than on outside perspectives in resolving his relational problems. Therefore, any weakened brand of relationalism that incorporates reference frames would seem largely incompatible with the supplementary criteria put forward in the *Principles*. A Cartesian may ultimately conclude, nonetheless, that a weakened formulation of relational motion, either (R1*) or (R2), is a more preferable state-of-affairs for the construction of a Cartesian spacetime than the supplementary criteria.

7.3. Critical Response to the Hypotheses.

In his important survey of these issues, Gueroult fails to notice the inherent deficiency in the WB displacement hypothesis which, as above, renders it incapable of ridding the Cartesian natural laws of relationalist obstacles. Rather than assess the criteria in this straightforward manner, i.e., on their ability to measure individual bodily motion, Gueroult ultimately judges their success against the backdrop of a deeper "definitional" puzzle that lies at the heart of Cartesian kinematics (which is the study of the motions of bodies regardless of force). In the *Principles,* Descartes stipulates that motion is the reciprocal translation of a body and its neighborhood, yet he then proceeds to stipulate that "by *one body,* or *one part of manner,* I here understand everything which is simultaneously transported (Pr II 24)." To avoid the intrinsic circularity of this approach to the phenomenon of motion, Gueroult reasons—correctly, it would seem—that Cartesian natural philosophy must presuppose an additional hypothesis that grounds the cohesion of the parts of extension (matter):

> What is required is a property that grounds—outside of all relativity [of the definitions, as above]—*displacement as a whole* [WB] and *single displacement* [SB]. This property is cohesion. Cohesion is due to the *rest of its parts,* as the *force* of rest, the force of resistance to the motion that would disaggregate them. If, thanks to this force, we can break the circle of relativity and reach the absolute of a science that is certain, it is because we have, by the same stroke, left purely geometrical [kinematical] concepts in order to raise ourselves to a dynamical concept—that of *force of resistance to motion*—which goes beyond these (212).

The "resistance to motion" manifest by Cartesian bodies, and implicitly at work throughout the collision rules, is thus singled out by Gueroult as the preferred solution to the "relativity", or circularity, of Descartes' definition of motion. Put differently, Descartes ultimately appeals to the dynamical aspects of motion to solve the riddle imposed by his haphazard kinematic descriptions (where "dynamics" refers to the motions of bodies under the action of forces). Gueroult's analysis of the dynamic factors implicated in Descartes' work may be correct (see chapter 6),[2] but it does not address the question of the potential success of these additional criteria in resolving our original problem of relational motion and the Cartesian natural laws. As presented in the *Principles,* these criteria were intended to provide a purely *kinematic* means of determining the individual motions of bodies for his physics. Moreover, the relational compatibility of Descartes' natural laws remains a serious problem *regardless* of the deeper circularity of the Cartesian definition of "motion" and "body." Gueroult does not critique this aspect of the criteria; therefore, after discussing some objections raised by Garber, we will need to return to the question of their overall success in answering the relational motion puzzle.

In response to Gueroult, Garber contends that the intended purpose or rationale of the supplementary criteria has been greatly misconstrued (1992, 346-348, fn. 12). As regards the SB displacement hypothesis, he argues that the context of the discussion strongly implies that it was invoked in order to explain our "common beliefs" about a reciprocal transfer between the earth and the object on its surface. In other words, the criterion was not designed to challenge the concept of reciprocal translation by procuring a means of determining which body is, or is not, in motion. Rather, the criterion merely constitutes a useful expedient in accounting for why we *think* that it is really the earth that is at rest. The "proper" view of motion, however, deems the translation to be reciprocal, so that attributions of rest or motion are purely relative once a motion occurs. Garber reasons that the supplementary criteria thereby serve as means of placating the Church censors (in the wake of the Galileo affair) so that he can advance his Copernican brand of planetary vortex theory without Church resistance.[3] Garber backs up this claim by recalling that Descartes prefaces his discussion of the criteria by stating that the reciprocity of transfer "clashes . . . with the common way of speaking (Pr II 29)," but later insists that "meanwhile we must remember that everything that is real and positive in moving bodies . . . is also found in the others contiguous with them (Pr II 30)." Descartes makes similar pronouncements with respect to the WB displacement hypothesis that appears in Pr III, where "the transfer [of the earth and the heavens] gives us no reason to attribute motion to the heavens rather than the Earth (Pr III 38)."

Garber's arguments are quite persuasive, but it is not clear that Descartes only intended his supplementary hypotheses to explain away our common perceptions of reciprocal translation. It is conceivable that the criteria both explain our perceptions of the phenomena, and, in addition, really do provide a means of determining the individual motions of bodies. In fact, in his exposition on planetary motion in Pr III, Descartes occasionally refers back to the criteria in a manner that suggest they are much more than a mere expedient to account for common misperceptions. For instance, directly after the quote from Pr III 38, cited by Garber above, Descartes declares:

> Moreover, in accordance with what was stated earlier, this motion [of earth and heaven] should only be attributed to the Earth; because the separation occurs over its whole surface, and does not similarly occur over the whole surface of the heaven but only over the concave portion contiguous to the Earth.

The reference to "what was stated earlier" concerns the crucial issue of the earth's proper motion, or lack thereof, as it sits in its band of circling plenum particles (that is, crucial for his relationship with the Church censors). In (Fr) Pr III 28, he invokes the WB displacement hypothesis to deal with this potential difficulty:

> In the common usage, all action by which a body travels from one place to another is often called motion; and in this sense of the term it can be said that the same thing is both simultaneously moved and not moved, depending on the different ways we determine its location. It follows from this that no movement, in the strict sense, is found in the Earth or in other Planets; because they are not transported from the vicinity of the parts of the heaven immediately contiguous to them, if we consider these parts of the heaven to be at rest. Since, to be transported in this way, they would have to be simultaneously separated from all the contiguous parts of the heaven, which does not happen. However, because the matter of the heaven is fluid, sometimes some of the particles, and sometimes other particles, move away from the Planet with which they are contiguous, and this is a movement which must be attributed solely to the particles and not the Planet. . . .

In this passage, Descartes contrasts the "common" with the "strict" sense of motion (as initially discussed in Pr II), and seems to argue that the reciprocal translation of the earth and some of its contiguous particles falls under the latter meaning of the term. The earth's lack of motion, and the motion of some of the contiguous particles, is motion in the "strict sense" because the earth is not, and a small number of particles are, transported from *all* of their contiguous neighbors. Consequently, at least in this passage, the WB displacement hypothesis is included within Descartes definition of the "strict sense" of the earth's motion. Needless to say, this passage (if interpreted correctly, here) would seem to contradict outright Garber's contention that the criteria were only meant to explain the "common sense" understanding of motion. Moreover, as is also evident in this context, Descartes needs both the SB and WB displacement hypotheses in order to satisfy the Church censors. Without a *literal* construal of these criteria, a motion between the earth and a body located on its surface could be legitimately ascribed to the former, and thus directly challenge the Church ban on terrestrial motion.

Furthermore, it is not the case, as Garber also contends (1992, 348), that the WB displacement hypothesis is only mentioned in Pr III of the *Principles*. Not only is it crucial for other sections in this Pr III, i.e., the analysis of the earth's motion in Pr III 28, as examined above, but it also appears at the beginning of his discussion of the SB hypothesis in Pr II: "we do not think a body is in motion unless it moves as a whole, and therefore we cannot understand that the whole earth moves simply since some of its parts are transported from the vicinity of some other smaller bodies which touch them; . . . (Pr II 30)." Indeed, the joint presence of the criteria in this Article is powerful evidence of their close interconnection. One might interpret the SB thesis, along these lines, as a particular instantiation of the more general WB hypothesis: since the two smaller moving bodies on the earth's surface exchange their entire contiguous neighborhood, while the stationary earth does not, the SB displacement hypothesis thus confirms or validates the broader WB displacement thesis. (Gueroult also notes the various interrelationships of these criteria; 1980, 210-212.)

It may still be objected, however, that Descartes' insistent reminder that motion is a reciprocal translation (of body and contiguous neighborhood) necessarily blocks any literal reading of the supplementary criteria. As Garber would undoubtedly point out, it is difficult to reconcile a strong brand of relationalism with a set of hypotheses which effectively breaks the symmetry of the relation. Yet, Descartes may have attempted to make the following, fairly subtle, distinction: although motion is a purely reciprocal *kinematic* translation, there are occasions when the "logic" of the kinematic events necessitates a breaking of the symmetrical translation, thus procuring individual states of bodily motion. In the SB displacement scenario, for instance, to ascribe motion to the earth is to ascribe a mutually *contradictory* state of affairs simultaneously to the same body. Generally, motion is a reciprocal translation, i.e., as a default view, but certain logically conflicting arrangements of transferring bodies can break the symmetry, as in the SB case. On this interpretation, Descartes' theory of motion would seem to resemble Leibniz' account, who also favored a purely relational, kinematic account of motion, but invoked non-kinematic elements to secure the determination of individual bodily motions. In the "Discourse on Metaphysics", he states:

> For considering only what it means narrowly and formally, that is, a change of place, motion is not something entirely real; when a number of bodies change their position with respect to each other, it is

> impossible, merely from a consideration of these changes, to determine to which of the bodies motion ought to be ascribed. . . . But the force or the immediate cause of these changes is something more real, and there is a sufficient basis for ascribing it to one body rather than another. This, therefore, is also the way to learn to which body the motion preferably belongs (Leibniz 1970, 315).

In a like manner, Descartes tried to procure a means of determining individual bodily motions (see, also, chapter 6). Yet, whereas Leibniz turned to an internal dynamic factor, Descartes introduced a set of uniquely kinematic criteria—a sort of "kinematic logic"—in order to break the symmetry of the reciprocal translation mandated for the success of his natural laws.

Overall, the above reconstruction has at least one key advantage over rival interpretations of the criteria: namely, it strives to accommodate both threads in Descartes' analysis of motion—the symmetry of translation *and* the criteria for determining individual bodily motions. The content of the discussion in Pr III 28 is of primary importance for this interpretation, needless to say, since it apparently contradicts Garber's claim that the supplementary criteria are not concerned with the proper, or strict, sense of motion (as change of neighborhood). Of course, the lack of any additional evidence supporting our construal of the criteria stands as a considerable obstacle to this interpretation. Yet, Descartes' neglect of this particular issue in his later work entails that Garber's reading (or any other reading, for that matter) suffers a similar fate.[4] All told, to accept Garber's conclusion that the criteria only concern our "vulgar" or common understanding of motion would seem to reduce Descartes' intricate hypotheses to the same sort of political expedient that earlier commentators, like Koyré (265) and Blackwell (227), attribute to Descartes' overall theory of relational motion—that is, the criteria are useful for placating the Church's ban on terrestrial motion, but they should not be taken seriously as integral components of Descartes' natural philosophy.

7.4. Final Assessment of the Hypotheses.

In the preceding section, we detailed the shortcomings of the previous interpretations of Descartes' supplementary criteria of bodily motion. However, we have yet to provide the final verdict on his ambitious attempt to delimit the individual motions of bodies (without violating the tenets of his strict relationalist reciprocity of transfer theory). By way of conclusion, therefore, we will turn once again to the problematic case of collision rules four and five in order to bring to light the strengths and weaknesses of the hypotheses.

As demonstrated, the WB displacement hypothesis fails to discern cases of type (i) from cases of type (ii), although the SB hypothesis can make this distinction. Nevertheless, as should be evident given the analysis in chapter 6, information on the individual motions of bodies is not sufficient in itself to rescue Cartesian physics from the clutches of underdetermination. To illustrate this point, once more, consider an example that Descartes provides to explain the third collision rule: "if [body] B had initially been traveling at six degrees of speed, and [body] C at four degrees of speed, both would subsequently move towards the left at five degrees of speed (Pr II 48)." Harmonizing relationalist doctrine with this sort of physical explanation is, indeed, a rather daunting task; but, for our purposes, attention should be focused upon Descartes' concept "degree of speed." All in all, it is difficult to see how the events represented in this impact rule, as in all the others, could be consistently explicated by appeal to anything less than a reference frame for measuring the bodies' speeds relative to one another. The determination of individual bodily motion by the SB displacement hypothesis is of no use, either; since, if the speed of the colliding bodies were a measure of reciprocal translation alone, then bizarre situations would occur wherein the approach speeds of bodies differed from their reciprocal translation speeds. For instance, returning to Descartes' example, although body B only approaches C at, say, one degree of speed, the "rate of neighborhood change" due to a joint or dual motion of both B and its neighborhood results in a local reciprocal translation speed of six (in order to match the example). Yet, as the context and explication of the collision rules clearly reveals, it is the common notion of an approach speed, and not the rate of translation of neighborhoods, that is upheld in the *Principles* (as also noted in chapter 6): for example, in the fourth rule, he stipulates that B could never move C "no matter how great the speed at which B might approach C (Pr II 49)."

Despite Descartes' best efforts, consequently, his supplementary criteria of motion cannot be reconciled to his laws of nature. What the natural laws require is a set of relationally consistent reference frames to ground the measure of individual bodily motions, a point which Garber also ultimately concedes: "without a common framework in which to conceive of the relative motions of more than one body, it is difficult to see how we could give an adequate treatment of the phenomenon of impact (171)." Yet, if this difficulty is acknowledged, then Garber's aforementioned claim that impact rules four and five are not inconsistent (241) is technically correct, but insufficient to save the Cartesian laws from a crippling relationalist deficiency. Essentially, the underlying intent or goal of Descartes' supplementary hypotheses, according to Garber, can be given the following reconstruction: because it is a fact that a body is, or is not, translating with respect to its contiguous neighborhood, there is a real distinction between the stationary earth and an object moving on its surface, or between rules four and five. This intuitively palatable conclusion thus resolves the difficulties associated with relational motion. Yet, even if a body at rest (not translating) is an invariant feature of Descartes' physics, the individual motion of the body is not an invariant once a translation occurs—and Cartesian physics mandates a consistent measure of these individual bodily motions (which is the aforementioned other needed half—see chapter 6). Put differently, even granting the reality of Cartesian motion, a relative motion is ambiguous as regards the individual components of motion (see chapter 1), and the natural laws *require* such individual determinations for their correct *application*.

Overall, it is this deficiency of Descartes' supplementary criteria—i.e., not providing a means of determining individual components of motion via some outside reference frame, via (R1*) or (R2)—that has not been properly diagnosed by previous commentators. Like Descartes' original, strict relationalist account of motion in the *Principles,* the supplementary criteria endorse an (R1) brand of relationalism, which as first diagnosed in chapter 6, are not sufficient to secure an adequate foundation for the Cartesian natural laws. Yet, as we have seen, a thorough examination of the criteria can reap the commentator many unexpected rewards. Not only do Descartes' criteria bear witness to his deep insights into the complex and problematic concept of relational motion, but they even anticipate the more successful strategies of later relationalists, like Leibniz. A host of latter-day relationalists can easily sympathize with Descartes' dilemma, and admire his attempted solution, even if his proposed kinematic means of determining individual bodily motions ultimately fails to ground his particular set of natural laws. In the next chapter, another committed early Cartesian, Huygens, will offer a more viable alternative for the construction of a Cartesian spacetime.

ENDNOTES

[1] A problem which comes immediately to mind is the general implausibility of requiring all moving bodies to manifest two oppositely-moving bodies upon their surfaces. Nevertheless, such seemingly extravagant demands are made in other places with respect to similar worries: i.e., the requirement that all bodily motions form part of some great circle of simultaneously moving bodies (in order to prevent the possible formation of a vacuum?); Pr II 33.

[2] Gueroult seems to hold that the "force of resistance" can be simply inferred from the speed of the striking body (1980, 202). Yet, as explained in chapter 2 and 6 above, "speed" is one of the relational, perspective-dependent properties of Descartes' physics (due to its obvious close connection to Cartesian "motion"); thus it would seem that the "rest force" should be seen as the more primitive notion, here, with speed the derived concept (via a reference frame "objectively" established by the postulated rest force—the invariant status of Cartesian "rest" providing the main work, of course). In short, the unique impact *behavior* of Cartesian bodies, which appears to invoke a resting force (but need not be interpreted as a *real* force inside the body), can serve as a basis for providing a consistent relationalist set of reference frames for measuring speed—once again, see chapters 5 and 6. As Argued in section 7.2, however, such reference frame procedures for rescuing Cartesian physics are problematic in that they seem incompatible with the strict (R1) relationalism implicit in the supplementary criteria.

[3] By defining motion as change of "neighborhood", Descartes could thus claim that the Earth remained at rest in its band of vortex particles (since the

contiguous particles did not change), as required by Church doctrine, yet still claim that this band *as a whole* moved around the sun. See, Pr III 28-29. See, also, chapter 3 and 6.

[4] As noted by Des Chene (1996, 266, fn. 11), Henry More believed that the discussion of the SB displacement hypothesis in Pr II 30 proved that Cartesian motion is not a symmetrical translation, and said as much in one of his many letters to Descartes (AT V 385). Unfortunately, and possibly of great significance, Descartes never provided a direct response to More's questions pertaining to this particular Article (i.e., Pr II 30).

CHAPTER 8

CONSTRUCTING A CARTESIAN DYNAMICS WITHOUT "FIXED" REFERENCE FRAMES: COLLISIONS IN THE CENTER-OF-MASS FRAME

As previously discussed, Newton singled out Descartes' espousal of an Aristotelian/Scholastic theory of place and motion as the primary target of his anti-relationalist assault. As mentioned in the Introduction, there would seem to exist two general strategies of countering Newton's allegations: (1) accept the contention that a fixed reference frame is incompatible with the tenets of spacetime relationalism (and attempt to provide an alternative foundation for Cartesian dynamics based on one of the laws of nature, such as the third law of motion); or (2) on the contrary, insist that such frames can be successfully and coherently established in a continuously changing matter-filled universe. This chapter, like the dynamics-based "rest" force procedure reviewed in chapter 6, will take the former option. That is, we shall investigate, once again, the feasibility of providing a framework for a consistent set of material-interaction laws *without* recourse to permanent reference frames. In this regard, Huygens' concept of a center-of-mass reference frame will prove invaluable, for it constitutes the a further means of preserving the basic content of Descartes' collision rules without jeopardizing the Cartesian conservation law (i.e., quantity of motion). In short, on this version of Cartesian science, we can one again retain quantity of motion and Descartes' theory of impact (from his third natural law). Unlike the method developed in chapter 6, however, Huygens' approach to the problem attempts to utilize the predictions of just one of the seven collision rules, namely, the first rule, as a means of reconstructing a relationally compatible version of all bodily impact.

Overall, we will demonstrate that a relationalist theory of space and time, when supplemented by Huygens' concept of a center-of-mass frame, can effectively accommodate many of the tenets of Cartesian dynamics. Nevertheless, we will also demonstrate that an augmented Cartesian theory still suffers from many of the defects that plagued Descartes' original hypothesis, and thus may ultimately prove to be an unsatisfactory attempt to resolve the difficulties inherent in Descartes' theory of motion and natural laws. In the process of examining the details of Huygens' theory, the context of our discussion will also necessitate an exploration of his various hypotheses on conserved quantities and centrifugal force. After the analysis of these theories, we will thus be in a better position to understand the intended purpose of Huygens' reconstruction of Cartesian dynamics.

8.1. Descartes, Huygens, and The Center-of-Mass Reference Frame

8.1.1. The Temporary Reference Frame Proposal.

As demonstrated in the first chapter, Newton's argument relies on the fact that the constant flux of the Cartesian plenum will preclude a fixed material frame or body from determining straight uniform motion (or velocity). Yet, Newton does not seem to envision the possibility of locating the reference frame alongside the moving or colliding bodies. Briefly, on this hypothesis, the uniform speeds and the straight paths of the two interacting objects (assuming we have such a case) are determined relative to a unique perspective or viewpoint that, in turn, is defined relative to the moving bodies. Therefore, since the bodies constitute their own reference frame, the need for a separately fixed material frame, or an absolute spatial position, is conveniently circumvented. That is, since both bodies move at constant speed *relative* to each other both before and after the collision, but not during the collision, of course, the entire event or system can be viewed as, what we will call, a "relatively-defined inertial frame." (The inherent problems of this tactic will be discussed in sections 8.3 and 8.4.) Naturally, a collection of such systems will not normally agree on the exact value of the quantity of motion with regard to one another. However, this fact does not necessarily preclude the possibility of a consistent Cartesian conservation law. If, for example, one were to modify Descartes' incorrect collision rules so as to accurately reflect experience from *within the relatively-defined inertial system,* then the Cartesian conservation law could be retained as legitimate physical principle. More precisely, a collision rule is required that, as viewed from a perspective inside the frame, actually preserves the total quantity of motion, which is defined as size times speed. (Alternatively, as developed in chapter 6, once can forego the need to reconstruct the collision rules and simply erect temporary reference frames as allowed in Descartes' handling of the original, un-reconstructed rules.) The scope of this law will be limited to each of these relatively-defined inertial frames, of course, because the total quantity of motion can only be preserved from a single perspective inside each system of colliding bodies. the total quantity of motion can only be preserved from within each system. This is a price a Cartesian may willingly decide to pay, nonetheless, for such a strategy would seem to reconcile Descartes' natural laws, which include his collision rules, with a relational theory of space and time. Finally, these frames must be located with, or "tied" to, the material occupants of the spacetime, so as not to invoke any form of absolute determination of motion, contra (R1), which a frame not linked to a material body would seem to entail.

8.1.2. Huygens and The Center-of-Mass Reference Frame.

One of the most successful attempt to rehabilitate Cartesian dynamics, along the lines suggested above, can be attributed to Descartes' younger contemporary, Christiaan Huygens. Huygens' scientific instincts, to a large extent, were entrenched in the amalgam of relationalism and vortex mechanics championed in Descartes' *Principles of Philosophy.* Like Descartes, he was inclined towards the incorporation of spacetime relationalism[1] with a contact-mechanical physics; i.e., with a theory that limits the interactions of material objects to direct body-to-body impact, and which denies the existence of action-at-a-distance forces, such as gravity.[2] His debt to Cartesian natural philosophy is obvious in the succeeding passage:

> To discover a cause of weight that is intelligible, it is necessary to investigate how weight can come about, while assuming the existence only of bodies made of one common matter in which one admits no quality or inclination to approach each other but solely different sizes, figures, and motions[3]

Even though he embraced many of his concepts, Huygens' work on mechanics far outdistanced the achievements of Descartes. Besides formulating several additional conservation laws of greater scope than quantity of motion—namely, the momentum law, mv, and *vis viva,* or mv^2(which roughly corresponds to the kinetic energy law)—Huygens' was also the first to provide a quantitative treatment of centrifugal force (see section 8.3).

In attempting to correct Descartes analysis of impact, Huygens selected as his starting point the first Cartesian collision rule, the only hypothesis in the entire set of seven actually verified through experimentation. This rule governs the collision of two equally sized bodies moving at the same speed (in opposite directions along the same line). Descartes asserts that both bodies will rebound along their initial path in the opposite direction "without having lost any of their speed (Pr II 46)," and thus conserve their total quantity of motion. Recognizing the importance of the first rule, Huygens sought to remedy the deficiencies in the Cartesian impact theory by utilizing an identical analysis for the remaining six cases treated in the *Principles*. By extending its scope over the collisions of all bodily sizes and speeds, the first rule can guarantee the conservation of quantity of motion for all interactions.

Huygens' search was motivated in large part by his understanding of the Principle of Galilean Relativity, which maintains that the state of inertial motion, or uniform speed in a given direction, does not affect the outcomes of physical processes. He realized that a single collision may appear, from one viewpoint, as a collision between a uniformly moving and a stationary object, but from a different perspective, as the collision of two uniformly moving bodies. His famous example involves an experiment with colliding spheres conducted on board a boat, which is sailing down river, as viewed from both the boat and the distant shore. Depending on the frame of reference, on the boat or on the shore, an observer will ascribe to each object an individual state of motion that will generally not agree with the observations conducted from another inertially related reference frame. For example, the observer on board the boat may judge that both objects are moving prior to their collision, while the observer on shore detects only the motion of a single body during the same time period. Like Galileo, Huygens therefore concluded that the state of inertial motion, or uniform speed in a given direction, does not affect the outcomes of physical processes:

> Thus we say that if the occupants of a boat travelling with uniform speed lets two equal spheres approach each other with speeds that are equal relative to the occupant and the parts of the ship then as a result of the collision each must spring back with speeds that, relative to the occupant, are exactly the same as if the occupant let the same spheres collide with the same speed when the ship is not travelling or he were on land.[4]

Consequently, the state of inertial motion of the bodies does not affect the outcome of the collision. With this in mind, Huygens endeavored to locate a frame of reference that would permit all bodily collisions to be viewed as a species of Descartes' first rule; i.e., where both bodies preserve their initial speeds (and hence quantity of motion) after rebound. His relationalist instincts, and his grasp of Galilean relativity, must have dictated that the outcome of all collisions should be identical regardless of the alleged motion and rest of the individual bodies involved. Nevertheless, the first collision rule only treats the interactions of equally sized bodies, a limitation of scope that greatly complicates Huygens' task. Although the relative speeds of bodies (i.e., their individual speed components) can be changed by simply adopting a different relatively non-accelerating frame, such transformations will not alter their relative sizes.

However, with the discovery of a colliding system's center-of-mass reference frame, which we will label a CM frame, Huygens found a means of generalizing Descartes' first collision rule to cover the interactions of various sized bodies. To grasp the significance of this concept, let us examine the case of two colliding bodies, labeled B and C, and assume that the latter has twice the size of the former, and that they both approach at the same speed.[5] If B is separated three feet from C, and we place a Cartesian coordinate system at a position one foot from C, then the products of their size and distance will be equal (2=2). Our presentation can be more precise, if x and y designate the respective coordinate positions of B and C, then the center-of-mass between these two bodies is the point where

$$xB = yC \text{ or } x/y = C/B \qquad (8.1)$$

(and the absolute values of these quantities are assumed throughout). In other words, viewed relative to this frame, the products of the size and position of our two bodies are equal. To maintain this perspective throughout the entire bodily interaction, consequently, it will be necessary to determine or locate the center-of-mass frame at each successive instant. On an absolute conception of space and time, the frame will thus need to constantly alter its position between the two colliding bodies to preserve the center-of-mass viewpoint (whereas, if equal-sized bodies approach at the same speed, the frame will remain stationary). However, since absolute states of motion are not sanctioned by the relationalist theory, the only verdict a Cartesian can provide is that, within the colliding system, a CM reference frame is continually specifiable, and that this frame may or may not be in a state of relative motion with respect to other bodies or reference frames. Given this judgment, the relationalist can claim that equation (8.1) is satisfied throughout a collision without having to classify the reference frame's individual state of motion relative to any outside systems.

Returning to Huygens' analysis, if we were to examine a collision from the CM frame, and employ Descartes' conservation law of quantity of motion, $Bv + Cw = Bv' + Cw'$ (where v and w signify the pre-collision inertial motion of B and C, and v' and w' denote their post-collision inertial motion) then, $v' = -v$ and $w' = -w$ (because, as viewed from that frame, both bodies merely reverse their direction after the collision). By substituting these results into Descartes' conservation law and simplifying both sides, we discover

$$v/w = -C/B. \tag{8.2}$$

An analysis of (8.1) and (8.2) reveals an important fact about the CM frame: Huygens deduces that "if a larger body A strikes a smaller body B, but the velocity of B is to the velocity of A reciprocally as the magnitude [size] A to B, then each will rebound with the same speed with which it came (*Oeuvres Complètes,* vol. 16, 92)." As viewed from the origin of that reference frame, where the bodies preserve their initial speeds after rebound, the ratio of their speeds is reciprocal to the ratio of their sizes.

With the disclosure of the CM frame, Huygens had thus found a relational means of conserving Descartes' quantity of motion in all types of collisions; but he had to reject most of Descartes' collision rules in the process, a realization that prompted him to assert: "If this [the CM frame] is granted, everything can be demonstrated. Descartes is forced to grant it however (*Oeuvres Complètes,* vol. 16, 96)." What Descartes is forced to grant is that six of his collision rules have been discarded, or refuted, by Huygens' CM hypothesis. When one body strikes another, irrespective of their size and speed, an observer situated at the CM perspective will perceive both bodies recoil in the opposite direction while retaining their initial speeds. In short, *relative* to this frame, all collisions are essentially the same. Moreover, it is important to remember that the total quantity of motion—defined as size times speed—is only faithfully conserved from the CM location: despite the fact that this frame will register the same amount both before and after the collision, other relatively non-accelerating (inertial) systems will generally reach different conclusions. From the perspective of most of these outside frames, the interaction of the bodies will fall within the scope of all seven Cartesian collision rules, which, as noted, are generally faulty. Since six of these seven rules will not provide accurate predictions, the total value of the quantity of motion will therefore not remain a conserved quantity across all such relatively non-accelerating frames. Once again, if we are attempting to uphold Descartes' conservation law in all material interactions, then the desired quantity, or invariant, is an identical value of the total quantity of motion *before and after* the collision, while the specific numerical value ascribed by each frame is arbitrary.[6] On Huygens' scheme, however, only one reference frame reliably accommodates this goal—the center-of-mass frame. From this perspective, the value of the quantity of motion will remain the same over the temporal period spanned by the movements of the bodies.

Prior to investigating Huygens' concept further, I would like at this point to digress from the main theme of this chapter (i.e., utilizing non-fixed reference frames to reconstruct Descartes' conservation law and collision rules) to explore some of the details of Huygens' additional conservation laws and his treatment of force. By focusing upon the development of these theories, we may gain an insight into the underlying reasons that prompted Huygens' strict adherence to a Cartesian natural philosophy, and hence his adoption of the center-of-mass reference frame. In short, Huygens is a fascinating, and often under appreciated, scientist of great genius who came very close to formulating the famous laws of motion which Newton later discovered on his own. Huygens' Cartesianism, in fact, is probably responsible for his failure to achieve the overt success which history has accorded to Newton. In the next section, consequently, we will examine Huygens' natural philosophy in an attempt better grasp the manifold purpose and effectiveness of Huygens' CM reconstruction of Descartes' collision rules.

8.2. Huygens on Conservation Laws, Impact, and Force

Ironically, even though he embraced Descartes' conservation principle, Huygens formulated several additional conservation laws of far greater scope, and was the first to provide a quantitative treatment of centrifugal force. In order to examine the underlying reasons for his continued acceptance of Descartes' conservation law, consequently, it will be necessary to elaborate the details of Huygens' further discoveries.

In many ways, Huygens' utilization of the center-of-mass frame owed much to his prior understanding of the principles of statics, the branch of mechanics that deals with bodies under the equilibrium of forces (e.g., the mechanical lever or pulley). In fact, our common experience of simple scales is probably sufficient to validate equation (8.1): if one wants to balance two unequal weights, then the larger weight must be positioned closer to the balance's point of suspension to bring about the desired equilibrium. The balance's suspension point is the center-of-mass, or, in this case, the center-of-gravity of the two bodies.[7] One of the most important principles of statics which Huygens' employed is that "the common centre of gravity of bodies cannot be raised by that motion of the bodies which is generated by the gravity of those bodies themselves (Huygens 1971, 154)." Based on this fact, as well as Galileo's work on free-fall, Huygens demonstrated that if two falling bodies reach speeds that are inversely proportional to their sizes, their collision must cause them to rebound with their original speeds (assuming they are perfectly elastic, or, as disclosed in chapter 4, perfectly hard). If not, the center-of-gravity of the two bodies will reach a height after collision that is greater than the height from which they fell, in direct violation of the above statics principle (and may result in a perpetual increase in motion—an accepted mechanical absurdity). By this means, Huygens' attempted to validate the Cartesian conservation law with the aid of his center-of-mass (-gravity) frame.[8]

Furthermore, by treating the possible vertical trajectories of bodies, as opposed to the merely horizontal, Huygens' thought-experiment introduced Galileo's uniform acceleration of free-fall into his analysis of motion (where the distance fallen is proportional to the square of the speed). This additional variable may be responsible for the genesis of a further law that we may call the conservation of kinetic energy. In a short paper submitted to the Royal Society in 1669, he declared that "the sum of the products of the size of each hard body multiplied by the square of its velocity is always the same before and after their collision."[9] (The "hard body" assumption will be discussed below.) Thus, let m_1 and m_2 represent the size of our two bodies, while v_1 and v_2 signify their pre-collision velocities (speed and direction) and u_1 and u_2 their post-collision velocities. Unlike Descartes' quantity, the product of the size and square of the velocity, to eventually become known as *vis viva,* is conserved in all (inertial) frames of reference both before and after the collision.

$$m_1 v_1^2 + m_2 v_2^2 = m_1 u_1^2 + m_2 u_2^2 \tag{8.3}$$

Huygens' paper likewise provides an early formulation of the law for the conservation of momentum, the same law that Newton would also advance. "The quantity of motion that two bodies possess may either increase or decrease as a result of their collision; but the quantity towards the same side, the quantity of motion in the contrary direction having been subtracted, always remains the same (Huygens 1989, 472)."

$$m_1 v_1 + m_2 v_2 = m_1 u_1 + m_2 u_2 \tag{8.4}$$

For both laws, the total quantity measured before the collision equals the total quantity measured after the collision, and this is true in all inertial frames. Descartes' law for the conservation of quantity of motion (QM), in retrospect, can be described as follows:

$$m_1 |v_1| + m_2 |v_2| = m_1 |u_1| + m_2 |u_2| \tag{8.5}$$

Here, the absolute value signs indicate the scalar value of Descartes' concept of speed.

For a relationalist, the scope of application of Huygens' new laws is much greater than his earlier CM notion, because the invariant quantities are conserved with respect to all relatively non-accelerating inertial frames, rather than a single perspective within each colliding system. Put simply, the extended scope of (8.3) and (8.4) is a basic fact of inertial reference frames (whether defined according to our relationalist procedure or not) and their respective coordinate transformation. In *all* inertial frames, the kinetic energy and momentum of the colliding bodies will be observed to remain the same before and after their collision.

Nevertheless, even after formulating these additional conservation laws, (8.3) and (8.4), Huygens' never completely abandoned his instinctive devotion to the Cartesian quantity of motion, nor did he forsake the notion of speed for the vectorial velocity concept. This adherence to orthodox Cartesianism is clearly manifest in his analysis of impact, an investigation that is mainly confined to the interactions of perfectly hard bodies (i.e., their shapes and internal constitutions do not deform or contort, even if only temporarily, under impact). He contends, in a late work dating from 1689, that soft bodies progressively come to rest in collisions, "but with hard bodies the situation is other, for their speed continues always without being interrupted or diminished, and therefore it is not surprising that they rebound (Huygens 1971, 155)." By his estimation, material impact completely preserves the speed of the moving bodies. More precisely, except for their reversal of direction, Huygens envisions the initial speeds as unchanged: his comment that the speeds continue "without being interrupted or diminished" seems to suggest that he does not view the collision as a process whereby the bodies regain or recover their initial speeds. This is an important distinction; for it vividly portrays what may be called a "kinematical" approach to material impact. For the Cartesians, the main emphasis of their collision treatises lies in the investigation of the pre- and post-collision speeds and sizes, which includes those quantities allegedly conserved. The physical processes that are involved in the actual impact of material bodies is a subject they do not discuss: that is, they do not focus their attention on the physical events that transpire in those brief instants when the bodies contact.[10] In a collision, the bodies merely reverse their directions and continue unabated, a purely kinematic phenomenon. To a large extent, this is also true of Newton's formal treatment of impact as regards the "coefficient of restitution."[11] Turning to Descartes, he likewise views material interactions mainly from the standpoint of the pre- and post-impact speeds needed to maintain a determinate conservation quantity (as observed in the Part II).

Huygens' remark also discloses the latent influence of Descartes' notion of speed. If the reversal of an object's direction does not change or alter its speed, then direction is not a defining property of this quantity. As with Descartes, Huygens separates the idea of speed from the idea of speed in a given direction: he elevates the former to a central position in his conservation laws while relegating the latter to a subsidiary function. Thus, despite his tentative experimentation with the concept of velocity in formulating two further conservation laws (8.3) and (8.4), Huygens shares with Descartes a fundamental scalar conception of motion. With respect to Huygens' conservation laws, Westfall points out that "the concept [of velocity] did not please him greatly, however, and he continued to speak formally of quantity of motion in the Cartesian sense, a scalar quantity which is always positive in value (Westfall 1971, 156)." Quite possibly, Huygens' professed Cartesianism is responsible for his continued adherence to the notion of speed, especially the influence of the relational theory of motion espoused in Descartes' *Principles*. If all determinations of motion are equally valid, then a conservation law based on the CM frame is no better or worse than a law founded on any other frame, despite the possible extended scope and applicability of the latter. A multitude of conservation laws thus follows as a natural consequence of the relational theory of motion, since all these laws employ reference frames that are ontologically alike. Nevertheless, there does not seem to exist any evidence that can pinpoint a deciding factor in Huygens' commitment to the notion of speed.

Huygens' relationalism does play an important role in his analysis of centrifugal force, however. Although the exact details of Huygens' discovery and quantitative formulation of centrifugal force are beyond the scope of this thesis, we can briefly review some of the his main conclusions. Like Descartes, Huygens sought to explain gravity in terms of the centrifugal force exerted by the rotating particles of the plenum (as outlined in chapter 3). He seemed to instinctively believe that the force exerted on a body in rotation is indiscernible from the force exerted on a body by gravitation. "For as all heavy bodies strive to fall with the same speed and acceleration, and as further this striving has a greater power the larger the bodies are, the same effect must hold for bodies fleeing from the center, whose striving is, . . . , entirely analogous to the striving that is caused by gravity (Huygens 1989, 489)." But unlike Descartes, Huygens was to quantify over these centrifugal effects and demonstrate that they follow the general form of Galileo's law for the acceleration of free-fall:

> If two bodies move around the same circle or wheels with unequal speeds, though uniformly, the ratio of the centrifugal force of the faster to the slower [which he had earlier identified with the distance of the radius connecting the circle to the tangent] is equal to the ratio of the squares of the velocities (Huygens 1989, 492).

In demonstrating that two accelerations, commonly held to be distinct phenomena, share a simple mathematical pattern, Huygens stood on the very threshold of the new physics of force. Yet, his ingrained Cartesianism prevented him from accepting the occult properties of "attraction" which proved so useful to Newton in unifying nature's effects. In 1669, he commented:

> To discover a cause of weight that is intelligible, it is necessary to investigate how weight can come about, while assuming the existence only of bodies made of one common matter in which one admits no quality or inclination to approach each other but solely different sizes, figures, and motions . . . (Huygens 1971, 186).

As noted in chapter 3, the one defining characteristic of Cartesian mechanics is the drive towards a purely kinematic force-less conception of body-to-body contact—a science that only necessitates the non-problematic observable properties of size and speed. We have examined the conflicting motivations that underlie Descartes' postulation of this form of theory; but Huygens' reasons are more mysterious. Not only did Huygens discover a set of conservation laws that seemed to supersede Descartes' quantity of motion, but he also was the first to correctly quantify centrifugal force effects. Nevertheless, these accomplishments did not trouble his instinctive Cartesianism, for he spent his entire career trying to salvage Descartes' contact-mechanical vortex (a theory we will examine in the next chapter). Yet, centrifugal force presents a different, and troubling case, for a relationalist, a point which will take up once again in a later section.

Finally, to unproblematically label Huygens a "Cartesian" would be somewhat disingenuous, however, for the details of his physics differed from Descartes' in several major respects. Most notably, Huygens felt it necessary to admit the existence of infinitely hard, impenetrable atoms separated by small empty spaces (vacuum), as well as a non-instantaneous velocity of light. As H. A. M. Snelders remarks: "In this acceptance of atoms and the intercorpuscular vacuum Huygens must be regarded as a Gassendian."[12] Despite these differences, Huygens invariably used Descartes as his starting point when theorizing about natural philosophy, as is evident in his investigation of bodily collisions.

In the next section, we will resume our discussion of the center-of-mass frame as we consider some of the Newtonian counter-replies to Huygens' method of preserving quantity of motion (according to Descartes' first collision rule).

8.3. The Newtonian Reply

Given our discussion in section 8.1, how would a spacetime "absolutist," who accepts the conclusion of Newton's *De gravitatione* argument, for instance, respond to the CM frame proposal? Besides invoking Newton's notorious "rotating bucket" experiment (see chapter 1), there are a number of objections that can be raised, some quite substantial.

8.3.1. Do CM frames violate Relationalism?

First and foremost, all substantivalists/absolutists will liken Huygens' frames to a covert reconstruction of the actual relationships and physical laws that obtain in absolute space. A Newtonian may ask: provided the relational equivalence of all frames in assessing motion, defined as (R1) above, why is a special class of coordinate systems (i.e., CM) singled out by Huygens' conservation law? The fact that the conservation laws do not hold from all perspectives, or even all inertial perspectives, will thus be interpreted as support for the spacetime absolutist view (although no absolutists of the seventeenth century appears to have presented this argument). In essence, they will insist that certain (inertial) reference frames are privileged due to the "embedding" of the laws of nature in the structure of absolute space and time, a fact that Huygens can never adequately explain due to the equivalence of all relational perspectives in determining motion. The spacetime "structure" that identifies these inertial trajectories for the absolutist (in modern Newtonian, or Neo-Newtonian, formulations) is the familiar covariant derivative, often labeled $\nabla_{\mathbf{V}}\mathbf{V}$ (or, $d^2x^\alpha/d\lambda^2 + \Gamma^\alpha_{\mu\nu}(dx^\mu/d\lambda)(dx^\nu/d\lambda)$, in the coordinate frame).[13]

Although it is not clear that the relationalist is committed to accepting some form of absolute space, the force of the absolutists allegations would nevertheless seem to demand a closer inspection of the particular brand of relationalism utilized by the CM frames: for, if only a special class of frames can uphold the conservation law, then the Cartesian spacetime would seem to possess a capacity, or structure, over and above the relations among bodies, in direct violation of (R1). Yet, like the dynamics-based "rest" force project employed in chapter 6, the Cartesian spacetime we are considering is a member of the class of Leibnizian spacetimes, which only possess a Euclidean spatial metric on the planes of simultaneity, as well as a time metric (see, Earman 1989, chap. 2). Hence, it is not the case that our CM frame method is built upon the much stronger $\nabla_{\mathbf{V}}\mathbf{V}$ structure of Newtonian spacetime. Rather, the Cartesian natural laws *themselves* serve as a sort of "bootstrapping" technique of locating the frames that conserve the desired Cartesian conserved quantities (likewise for the dynamics-based "rest" force process in chapter 6). On this explanation, consequently, it is the sparse structure of the Cartesian spacetime in conjunction with the Cartesian natural laws that pick out the privileged reference frames. Overall, the CM frames would appear to play a role similar to that of the "fixed stars" in Mach's attempt to resolve Newton's "bucket" experiment. In order to account for the centrifugal effects of the water's rotation, Mach postulated that the force effects were not due to the water's acceleration relative to absolute space, but a result of its acceleration relative to the fixed stars. Mach's strategy, which (at least hypothetically) eliminates the need for absolute space, is succinctly summarized by Sklar: "where Newton fails, [Mach] argues, is in his attempt to show that no material object could be the proper reference frames for the absolute accelerations to be relative to."[14] Mach's fixed stars and the CM frames both constitute a relationalist means of choosing a class of materially-based inertial frames from which to explicate material phenomena (rotation and collisions, respectively)—but, the relational spacetime structures are weaker than $\nabla_{\mathbf{V}}\mathbf{V}$ because they cannot ascertain the inertial continuation of the privileged material-based frames independent of the particular physical laws, either Descartes' or Newton's, coupled to those worlds.

Since the CM method essentially follows the dynamics-based "rest" force strategy in so far as it adopts a weaker formulation of relationalism, we would appear to have another instance of an (R1*) relationalism, which can (once again) be defined as:

R1*: All motion is the relative motion of bodies, and consequently, spacetime does not have, and cannot have, structures that support "absolute" quantities of motion. But, a privileged class of reference frames may be adapted to the invariant quantities of motion of the spacetime as long as it does not entail a fixed, absolute spacetime structure.

The invariant quantities of motion in this Cartesian spacetime are the "relative body speeds (velocities, accelerations)," which fail to qualify as "absolute" since they are only determined relatively among bodies and not to absolute space. Given the invariant quantities of this relational (Leibnizian) spacetime (as above), and the Cartesian conservation law, the plan is that the needed CM frames can be chosen without invoking $\nabla_V \mathbf{V}$. Overall, this form of relationalism has the advantage of nicely accommodating Descartes' actual treatment of motion, which, as we have seen, violates (R1), alongside his repeated denials that space and time (and hence motion) are anything more than relations among bodies: e.g., "the names 'place' or 'space' do not signify a thing different from the body which is said to be in the place (Pr II 13; see also, Pr II 8-18)." Unfortunately, as also noted in chapter 6, it is unclear if (R1*) is generally successful, or even consistent, as there is a sneaking suspicion that an underlying Newtonian structure is being dressed in a relationalist guise. In a sense, the critic may charge that the conjunction of Descartes' natural laws and the invariants of its spacetime simply provides the same structure as Newtonian spacetime, but merely relabeled to suit relationalist sensibilities. Throughout the remainder of this chapter, however, we will assume that (R1*), and/or (R2), to be discussed below, are relationally consistent.

In its favor, though, the major difficulties for a relationalist maneuver such as (R1*) would seem to have been historically associated with the phenomenon of rotational motion (Newton's "bucket", again),[15] and not with the sorts of collision hypotheses upon which our CM frames, or dynamics-based "rest" force frames, have been constructed. Rotation, in contrast to the type of bodily collisions viewed from a CM or "rest" force frame, presents grave empirical problems for a relationalist (and likewise extols the virtues of $\nabla_V \mathbf{V}$-equipped spacetimes). In the Machian theory described above, the rotation of the fixed stars relative to a fixed bucket is a relationally indistinguishable state-of-affairs from a rotating bucket and fixed stars. Yet, Newton's theory, as well as the General Theory of Relativity, predict observationally distinguishable effects for these two cases (i.e., the centrifugal force effects), which suggests that Newtonian and General Relativistic dynamics are not congenial to Mach's version of (R1*). Our CM and "rest" force frame rendition of Cartesian dynamics, however, since they treat a limited class of bodily collisions in a *kinematic* fashion, are thus freed from the potentially disastrous need to explicate the more complex non-inertial phenomena manifest in bodily rotations (or other forms of acceleration, e.g., motion through a non-uniform gravitational field). The kinematic approach to collisions evident in the work of Descartes and Huygens, where bodies merely approach along a straight line at uniform speeds and reverse their direction after impact (while ignoring the processes that occur during impact), thus constitutes a more suitable phenomenon for a marriage of (R1*) and a set of inertial laws of motion than Mach's similar conjunction of natural laws and (R1*) for the more dynamically sophisticated behavior of bodily rotation. Mach tried to explain rotation in an equally kinematic fashion, as simply movement relative to a material reference frame, but the dynamic behavior of bodies undergoing these motions breaks the symmetry of this kinematical approach.[16]

Huygens' discovery of the correct quantitative formulation of centrifugal force nicely demonstrates the troublesome questions that such dynamical phenomena can raise for a devoted relationalist. Unlike his conservation laws for quantity of motion, momentum, or *vis viva,* a body's centrifugal force cannot be removed by simply adopting a new reference frame. To put it differently, the quantities conserved in the interactions covered by the above three laws do not hold in all possible frames (i.e., transformations to non-CM or non-"rest" force frames for quantity of motion, and transformations to non-inertial frames in the case of the other laws, will generally fail to conserve these respective quantities). As Westfall has suggested, and Gabbey's work seems to imply,[17] this fact may have contributed to Huygens' continued interest in Descartes' quantity of motion even after his discovery of the more general momentum and *vis viva* laws, since the "forces" apparently conserved in all these laws could be regarded as a mere manifestation from a particular point of view—and with many possible points of view, there may be different quantities conserved in each different reference frame. However, (assuming Mach was wrong), no matter what perspective one takes with respect to a relative rotational motion among two bodies (i.e., which body is considered to be at rest and which body is rotating), it will still be the case that only one body, the rotating body, will experience the centrifugal force. The existence of this force would seem to strongly imply that at least some quantitative force phenomena are *independent* of the particular reference frame selected to view the behavior of bodies, a conclusion that would seem to challenge the "kinematical" approach to bodily collisions apparent in Huygens' work, such as *De motu corporum ex percussione* (and Descartes' collision rules, Pr II 46-52). Specifically, one cannot eliminate or discharge the unwanted behavior of bodies by merely shifting to a different reference frame.

Finally, a relationalist unsatisfied with the "bootstrapping" process implied in (R1*) may want to once again fall back on the relational strategy (R2) instead, which, once again, regards spatiotemporal relations among bodies as direct; that is, they are not parasitic on relations to a absolute/substantival space that underlies bodies. This form of reasoning is the relationalist's last line of defense, for it is merely the denial of spacetime "substantivalism" (i.e., spacetime as a substance or entity that is independent of, or can exist apart from, matter). If (R2) is accepted in place of (R1*), consequently, then the relationalist is ostensibly free to invoke whatever spacetime structure is deemed necessary to, in our case, conserve the Cartesian natural laws (possibly even $\nabla_{\mathbf{V}}\mathbf{V}$). As long as the existence of these richer structures are regarded as somehow contingent upon the material occupants of the plenum (however that may be), the relationalist can seemingly lay claim to the Newtonian legacy without the added ontological commitments. But the real question, once again, is whether or not (R2) is a consistent relationalist alternative to (R1), or even (R1*); a verdict, furthermore, which is still pending (see note 13).

8.3.2. The Physical Scope of the CM Frames.

Our second absolutist counter-reply to the CM frame method is a follow up to the first: in short, if the CM frames are spared the difficult task of determining rotational (accelerated) motions, then this formulation of Cartesian physics will necessarily fail to up hold the conservation law for all material events. If the CM frames are to be used to preserve the quantity of motion in all physical processes, and thus satisfy Descartes' demand for an overall conserved universal quantity, then it will be necessary to view *all material interactions* as a form of collision subsumed under the first collision rule. Nevertheless, as described, Huygens' reconstruction of the first collision rule is confined to a distinct class of material interactions, namely the impact of bodies moving along straight paths at uniform speeds relative to one another. This is all the more problematic when one discovers the substantial role that (non-uniformly) accelerated motions assume in the Cartesian plenum. To give one example, the plenum particles that constitute a large ring of circling matter, or vortex, will momentarily increase their speeds when passing a narrow channel or obstruction situated along their path. In such cases, Descartes insists that:

> all the inequalities of the spaces [of the circular path] can be compensated for by corresponding inequalities in the speed of the parts. So, without there being any condensation or vacuum; all the matter contained in the space EFGH can move in a circle. The part of it which is near E can move toward G and that which is near G can simultaneously move toward G, provided only that (since we are supposing the space at G to be four times as wide as at E, and twice as wide as F and H) we also suppose the movement to be four times as rapid at E as at G, and twice as rapid as at F and H. Similarly, in all remaining places, we can suppose that speed of movement compensates for narrowness of space. Thus, in any given length of time, the same quantity of matter will pass through one section of the circle as through another (Fr Pr II 33—see Figure 23).

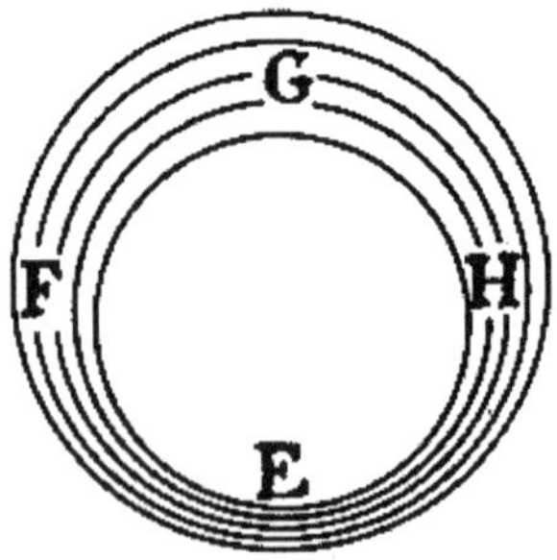

Figure 23. Descartes' illustration from the *Principles,* Pr II 33.

Given this scenario, one may wonder if it is therefore possible to comprehensively explain all natural phenomena as a form of interaction characterized by uniform pre- and post-impact relative speeds. Since the vortex particles variably accelerate to compensate for obstructions along their path, their random collisions will not normally exhibit the relative constant speeds required to apply effectively the method of the CM frame, and hence conserve quantity of motion. This criticism, of course, also applies to the dynamic-based "rest" force procedures outlined in chapter 6, since the temporary frames installed were similarly confined to impacting bodies exhibiting constant speeds. (Once more, there does not appear to be any historical precedents for this argument against the CM frame.)

8.3.3. Predictive Scope and the CM Frames.

Our third absolutist response closely resembles the preceding argument. If the CM frames suffer from a limitation of scope with respect to the types of collisions they can successfully explicate, they seem further restricted by an inability to predict or determine the future states of material bodies after they have departed the frame. Specifically, since a CM frame only measures a given body's quantity of motion during the brief periods spanned by its impact with a second body, one might conclude that it apparently cannot offer any predictions of the future states of this quantity after the bodies have separated and joined in other collisions. Each particular CM frame, consequently, can only track a body's motion, and determine its product of speed and size, for specific finite (infinitesimally small) temporal intervals. If placed within the confines of the Cartesian plenum, *where bodies constantly collide,* the predictive scope of Huygens' frames is subsequently restricted to mere instants. During even the briefest of intervals, a given object will be engaged in a vast number of distinct collisions with a host of adjacent bodies, a situation that would most certainly limit the determination of a body's future motions to an equally short period of time, if not single instants. In addition, there is the further problem of how an infinity of nearby, possibly simultaneous, collisions affects Huygens' method, since such a possibility would seem to greatly complicate, if not hinder, the application of the CM frames.

Summarizing the third absolutist argument: at best, the CM theory can only provide a measurement of the quantity of motion at each separate or distinct collision, but not continuously over a series of such interactions. A Cartesian would be disinclined to accept this judgment if it entailed a restriction to mere instants, however; for such a confinement of this quantity would appear to conflict with Descartes' analysis of motion. Overall, Descartes envisions motion as a process that necessarily involves a temporal duration, for "no movement is accomplished in an instant (Pr II 39)." Since quantity of motion employs speed, limiting the Cartesian conservation law to single instants would therefore likely raise serious textual objections. In addition, a Newtonian would probably insist that this method of determining motion runs counter to our normal measuring procedures, if not common sense intuition. An object does not require a new reference frame as it approaches every fresh contact; rather, its motion, and hence quantity of motion, can be continuously traced over any given number of collisions. The absolutist will accordingly interpret our common experience of impact measurements as support for the existence of absolute space, since there must be some mechanism or medium that permits the continual estimation of these physical quantities.

The problems engendered by the third absolutist counter-argument are not necessarily disastrous to the CM frame project, however. In the next section, we shall demonstrate how a CM frame, when carefully specified, can attempt to overcome the limitations of its inherent lack of predictive scope.

8.4. Constructing a Center-of-Mass Reference Frame

How should the Cartesian respond to the allegations of the third Newtonian argument? Obviously, if the CM frames are to be retained, the relationalist will need to procure a means of determining and preserving a body's quantity of motion over the course of several collisions. That is, a procedure must be obtained that will permit the coupling or linking of each distinct CM coordinate system (reference frame), and hence allow information on the status of bodies in one frame to be inferred from another frame. The problem, as described above, is based upon the fact that all bodies eventually enter collision frames that are apparently not related to their previous collision system. We can detail the argument as follows: (1) since we are assuming a relational theory of spacetime, each center-of-mass coordinate system is in a state of relative motion with respect to all other systems; thus, their exist no meaningful non-relative or individual determinations of a frame's state of motion (which is true regardless of whether we accept an (R1) or (R1*) construal of the CM project). (2) In order to conserve a body's quantity of motion, the determination of this quantity must be conducted from the CM frame; hence, the particular value assigned to each colliding body is fully dependent upon that coordinate system. Consequently, as is evident from the conjunction of statements (1) and (2), the value of an object's quantity of motion is entirely relative to, and thus only *meaningful in,* its current CM system.

Nevertheless, this argument overlooks a significant fact concerning Huygens' frames: although they do not display individual non-relative states of motion, one can determine the relative speed and distance between two or more frames (as noted above). As a consequence of the spacetime structure, a Cartesian can now meaningfully and coherently discuss the "relative speed (or velocity, acceleration, etc.)" among several bodies without violating the tenets of relationalism. These quantities are invariants of our relational (R1), or (R1*), spacetime, and is thus a member of the larger class of Leibnizian spacetimes (see chapter 2). Likewise, all reference frames will calculate an identical spatial difference between bodies, as well as an equal measurement of size. Therefore, despite its sparse appearance, our Cartesian spacetime exhibits a number of invariant properties: all frames will observe the same bodily sizes, relative distances, and relative speeds among bodies.

To demonstrate a relationalist means of correlating the CM frames, consider the following example: suppose a material body with size B departs the origin of its CM frame F_b with a speed v (measured with respect to F_b), while a second body C exits the center point of its CM frame F_c with relative speed w. Furthermore, assume that B and C approach one another on a collision course directed along the same straight line. Given this scenario, the absolutist will correctly infer that a third CM frame $\overline{F}$ is required to measure the quantity of motion of the soon-to-collide B and C (see Figure 24); but, they will also conclude that their respective speeds $\overline{v}$ and $\overline{w}$, which are determined relative to $\overline{F}$, will be different from the previously assigned values v and w (relative to F_b and F_c). Due to the relative motion of F_b, F_c, and $\overline{F}$, the speeds ascribed to the two moving bodies will not normally be the same. However, although it is true that the value of a body's speed is frame-dependent, this does not exclude the possibility of predicting the magnitude of this quantity in several distinct frames based on the observations conducted from a single frame. More specifically, Cartesian spacetime allows an observer located at either F_b or F_c to determine the exact values of $\overline{v}$ and $\overline{w}$ *in* the CM coordinate system $\overline{F}$.

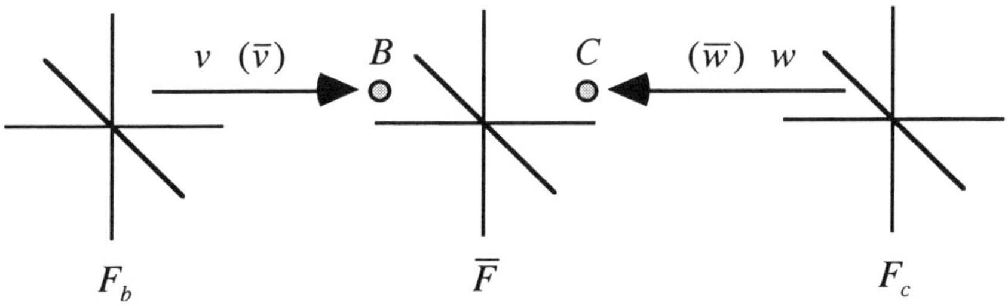

Figure 24. The objects departing the CM frames F_b and F_c, with speeds v and w respectively, will collide in the CM frame $\overline{F}$ (with speeds $\overline{v}$ and $\overline{w}$).

The following example can establish this point: due to the invariance of relative speed, assume that the frames F_b, F_c, both calculate a speed difference between B and C of, say, 6 miles per hour at time t_0. Also, from the invariance of mass and distance relations, suppose our frames measure a spatial difference of 5 feet at t_0, and that the respective sizes of C and B are as 2 to 3 (i.e., $C/B = 2/3$). Given these numbers, it is easy to predict the precise values that $\overline{v}$ and $\overline{w}$ will take in $\overline{F}$, which is the CM frame of the impending collision between B and C (which will transpire at a time later than t_0). If we recall Descartes' conservation law (8.5) and

(8.1)-(8.2), and apply them to $\overline{F}$, we arrive at the equation (assuming absolute, i.e., positive, values)

$$C/B = \overline{v}/\overline{w} = \overline{x}/\overline{y} \tag{8.6}$$

where $\overline{x}$ and $\overline{y}$ are the respective coordinate positions of C and B in $\overline{F}$. Thus, since we know that $C/B = 2/3$, frames F_b and F_c will both determine at t_0, when the invariant spatial difference is 5 feet, that the origin of $\overline{F}$ should be located at a position where $\overline{x} = 2$ and $\overline{y} = 3$ (relative to $\overline{F}$, of course). In addition, the invariant relative speed is 6 m.p.h.; thus, F_b and F_c will both predict that $\overline{F}$ will assign the speed values $\overline{v} = 2.4$ and $\overline{w} = 3.6$ (since their ratio is 2/3 and their sum 6). In conclusion, the frames F_b and F_c can both determine at time t_0 the exact placement of the origin of $\overline{F}$, and the speeds that $\overline{F}$ will gauge for C and B. Once we are provided these magnitudes, the quantity of motion as measured from $\overline{F}$ is obtained by multiplying the relevant speed and mass (or Cartesian size) values: here, we should also note that (2.4)3=(3.6)2 as required by equations (8.1-8.2) and the Cartesian conservation law (8.5).

Of course, as the bodies draw closer together over time, the values obtained for $\overline{F}$ from both F_b and F_c will change; that is, this method will determine new values at each succeeding instant. Yet, since the bodies in Huygens' frames move at relative uniform speeds before and after impact, so that all the frames are inertial relative to one another, our observers at F_b and F_c can easily estimate the precise spatial location where C and B will collide, say, at a time t_n, and the exact location of the CM frame $\overline{F}$ for each instant from t_0 to t_n. That is, because the bodies move inertial relative to one another, their relative velocity will remain invariant; so, $\overline{v}$ and $\overline{w}$ will retain the same value throughout the time period (only their coordinate positions $\overline{x}$ and $\overline{y}$ will change in order to preserve the ratio 2/3). Consequently, as was initially desired, we have located a means of connecting or correlating the value of the quantity of motion of a body (C or B) in one frame (F_b or F_c) with its quantity of motion in another frame ($\overline{F}$) over a future temporal interval (t_0 to t_n). Although this relationalist procedure needed both C and B to derive the correct total results, once these values are obtained we can easily infer each object's individual quantity of motion. In short, a Cartesian can now employ Huygens' frames to predict the future coordinate systems that preserve Descartes' quantity of motion, along with the values ascribed to the individual bodies relative to those future frames. By demonstrating an ability to compare quantities across frames, this relationalist strategy would thus appear to overcome at least one important absolutist objection. Finally, it should be noted that a similar construction can be provided for our previous dynamics-based "rest" force procedure for preserving quantity of motion.

However, it is important to note that this method of conserving Descartes' quantity of motion is not a conservation law in, what we may call, the standard or "classical" sense of the term. The classical conservation laws can track a particle's motion and conserve the desired quantity from the perspective of a *single* (inertial) frame, through employing such conservation principles as momentum or kinetic energy. Yet, our Huygens' system can, at best, only offer predictions on the positions of the future CM frames that will conserve quantity of motion *from the perspective of that future frame*. For example, on Huygens' scheme, a center-of-mass frame, say, F_b, can only predict the location of another CM frame, such as, $\overline{F}$, that maintains the value of the desired invariant from $\overline{F}$'s viewpoint: F_b cannot continuously track a body's motion and calculate the law's invariant quantity over a succession of interactions from the perspective of F_b, as is possible with the standard conservation laws. This is a significant realization, but is not necessarily a major problem for a Cartesian intent on utilizing Huygens' method of conserving quantity of motion. Overall, Descartes' principle that "[God] always maintains an equal quantity [of motion] in the universe (Pr II 36)," is subject to many interpretations; and it is certainly not clear that he had in mind the classical conception of a conservation law. A conservation law that relies on a series of momentary frames defined relative to individual collisions, rather than from a single frame that covers a series of collisions, would seem perfectly compatible with Descartes' request for a relationally-defined conserved quantity of motion. This last point corroborates our conclusions on the quasi-holistic character of the Cartesian conservation principle that was revealed in chapter 5: i.e., since Descartes did not envision his conservation law in the modern sense, as a local quantity that is conserved from the perspective of all inertial reference frames, it is possible to expand the allowable interpretations of "conservation principle" to incorporate Descartes' unique contribution.

With respect to this method of linking the CM frames, finally, an absolutist might raise the objection that this process seems to bestow a sort of global inertial frame upon our Cartesian spacetime, and thus constitutes a violation of relationalism by tacitly acknowledging more structure than the relationalist can admit. There are two replies to this argument, which correspond to the two alternative relationalist variants discussed in section 8.3. First, if (R2) is employed, then a global inertial frame would not represent a violation of relationalism, as long as the frame was dependent upon the material occupant of the universe (whatever that may mean). Second, according to (R1*), or even the original (R1), our method of linking the CM frames does not appear to pose a threat to relationalism since (as we have seen) this process *cannot* track a single particles' motion over an indefinitely long temporal period, a capacity seemingly possessed by "real" global inertial frames. On the whole, given the persistent collisions in Descartes' plenum, the relatively brief life span of each CM frame adapted to the momentary invariants of Cartesian spacetime does not seem to admit the possibility of predicting future CM frames for more than a few fleeting instants.

8.5. Conclusions

We began this chapter with a particular question in mind: Is it possible to harmonize Descartes' relationalism, conservation law, and hypotheses on bodily interactions without assuming the sort of absolute or "fixed" reference frames assumed necessary by Newton's *De gravitatione* argument (as implicated in premise (5) of the argument; see chapter 2)? After much consideration, the answer is yes, but with serious reservations. Although Huygens' project is handicapped in many areas, the Newtonian must still admit that Descartes' quantity of motion is conserved in collisions viewed from the center-of-mass perspective. More importantly, this conservation law functions within a relational spacetime, and thus obviates the need for Newton's absolute space, or at least a substantivalist interpretation of Newtonian spacetime, to determine a body's trajectory. Of course, six out of seven of Descartes' impact rules were abandoned along the way, and the form of relationalism advocated by the Cartesian changed from (R1) to either (R1*) or (R2), but the remaining collision rule serves as the foundation for a working Cartesian dynamics (acknowledging, once again, the ambiguity of "Cartesian dynamics" as a descriptive term). In addition, as remarked in section 8.3 (and chapter 6), (R1*), or (R2), appear to be a more accurate reflection of Descartes' actual treatment of motion, regardless of his somewhat deceptive avowal of (R1).

Yet, despite the development of a means of linking the information across distinct CM frames (section 8.4), there still exists powerful reservations connected with Huygens' scheme: even if we grant the disputed consistency of (R1*) and/or (R2), the CM reference frames apparently cannot explicate the interactions of (relatively) accelerating bodies. As noted, if one attempts to conserve the total quantity of motion throughout the universe, as mandated in Descartes' *Principles,* then it will be necessary to provide CM frames for the collisions of bodies not exhibiting uniform pre- and post-impact relative speeds, since these interactions are rather prevalent in Descartes' world. Nonetheless, quantity of motion will not normally be conserved from a CM perspective in collisions involving such accelerating bodies. If these frames cannot conserve the Cartesian universal quantity, then the method for linking the CM frames outlined in section 8.4 will be to no avail. Our dynamics-based "rest" force approach, from chapter 6, suffers from this same inability to cover the interactions of accelerating bodies. It is important to bear in mind that the limited scope of the Cartesian conservation law is an obstacle for a consistent Cartesian physics, and not necessarily for the other conservation laws developed in the seventeenth century, such as (8.3) or (8.4), since Descartes was insistent that his law governed the interactions of all material phenomena, including (presumably) gravitational free-fall (Pr II 36). Leibniz was quick to spot this deficiency in Descartes' theory (see chapter 5), and offered *vis viva* as the correct universal conservation principle; while Newton, on the other hand, largely confined his analysis of the conservation of momentum bodily interactions on flat surfaces, which thereby cancels out the effects of gravity. However, on the bright side, a committed Cartesian relationalist can at least claim the quantity of motion is consistently conserved within the context of the collision rules—and these seven types of material interactions are of fundamental importance since they comprise the only worked out dynamical phenomena in the Cartesian corpus.

ENDNOTES

[1] Huygens' relationalism seems to have remained largely (R1), as is evident in the following passage: "there is nothing to distinguish straight motion from rest, and both one and the other are relative. . . ." Huygens (1950, vol. 16, 183), trans. by A. Elzinga (1972, 96). But, the center-of-mass reference frame seems to usher him into the realm of (R1*) relationalism, as will be discussed below. See, also, endnote 2.

[2] For a discussion of Huygens' role in the history of relational spacetime theories, see Stein (1977). Stein suggests that Huygens' treatment of rotation favors a much weaker commitment to relationalism than has normally been acknowledged by commentators: that is, using our classification scheme, Stein holds that Huygens' theory of rotation would fall under (R2), or (R1*), and not (R1). However, I concur with Earman (1989, 67-71), who regards Huygens' brief statements on rotation as too vague and underdeveloped to constitute a clear instance of (R2)/(R1*) relationalism. For various aspects of Huygens' natural philosophy, see Elzinga (1958) and Dugas (1958, chap. 10).

[3] Huygens (1950, vol. 16, 186), trans. by Westfall (1971, 149). Unless otherwise noted, all following translations will be based on Westfall, but checked against the *Oeuvres Complètes.*

[4] C. Huygens, "De Motu Corporum ex Percussione," in *Oeuvres Completes,* vol. 16 (1929, written 1656), 31; trans. R. J. Blackwell in *Isis,* 68, 574-597 (1977).

[5] I owe many of the details of the following discussion to Barbour (1989, 473-478).

[6] In this context, I am using "invariant" to signify a quantity of a physical law that retains the same value over time. Recalling our discussion of differential geometry from chapter 2, 'invariant' can also be used to describe a quantity of a physical theory that retains the same value in all the reference

frames formed by an admissible coordinate transformation (i.e., the covariance group of the theory).

[7] In addition, it is important to note the possible influence of the "virtual velocity" tradition in Huygens' decision to use of the center-of-mass frame to conserve quantity of motion. On this theory, the equilibrium of suspended weights is caused by the balancing-out of their products of downward-directed speed, possibly as some sort of potential speed equivalent to weight, and distance from the center point. This view is nicely presented in one of Galileo's lesser known works: "Two weights equall in absolute Gravity, being put into a Ballance of equall Arms, they stand in Equilibrium, neither one going down, nor the other up: because the equality of the Distances of both, from the Centre on which the Ballance is supported, and about which it moves, causeth that those weights, the said Ballance moving, shall in the same Time move equall Spaces, that is, shall move with equall Velocity, so that there is no reason for which this Weight should descend more than that, or that more than this; and therefore they make an Equilibrium, and their Moments continue of semblable and equall Vertue." *Discourse on Bodies in Water,* trans. T. Salusbury, ed. S. Drake (Urbana: University of Illinois Press, 1960) 6-7. Rohault's important Cartesian text also presents this notion. See, *A System of Natural Philosophy, vol. 1* (written 1671), trans. J. Clarke and S. Clarke (1723), (New York: Johnson Reprint Corp., 1969), 43-44. However, it is not known if the virtual velocity concept actually played a role in Huygens' approach to quantities conserved in collisions, a set of circumstances somewhat different than wieghts held in equilibrium. For more on virtual velocities, see chapter 5.

[8] This proof appeared in 1656, four years after his initial discovery of the center-of-mass method for preserving quantity of motion. See Westfall 1971, 148-159.

[9] C. Huygens, *Oeuvres Completes,* vol. 16, 233; trans. J. B. Barbour, 1989, ibid., 474.

[10] This conceptualization of impact is brought out in precise detail by Rohault, an influential Cartesian of the later seventeenth century. With respect to a body whose direction is reversed after impact, he states: "Because the Notion we have of *reflected Motion* is not different from the Notion we have of *direct Motion,* we ought not to think that these Motions are contrary to each other, but that the one is only a Continuation of the other, and consequently, that there is not any Moment of Rest in the point of Reflection, ... Besides, if a body which was in Motion, comes to be but one Moment at Rest, it will have wholly changed its manner of existing into the contrary, in which there will be as much Reason for its continuing, as if it had been at Rest a whole Age; . . . ." To these views, Smith's footnotes offer a Newtonian reply: "There may be a Moment of Rest, in the point of Reflection; because the reflected Motion, is not a Continuation of the Direct;

but a new Motion impressed by a new Force, *viz.* the Force of *Elasticity.*" J. Rohault 1969, 81.

[11] Despite Newton's third law and his acknowledgement of elastic force, it must be admitted that his analysis of imperfectly elastic collisions (where the bodies do not recover their initial speeds) displays a strong kinematical bias, since he merely compares the pre-impact and post-impact velocities to determine their relative ratio. If the bodies are perfectly elastic or perfectly inelastic, this ratio will equal 1 and 0 respectively (i.e., in the latter case, the bodies become attached during impact). For all bodies with an internal constitution in-between these values, a factor k $(0 \leq k \leq 1)$ is needed, the "coefficient of restitution," to convert the post-impact velocity to the value that obtained before the collision. Hence, Newton's measurement of the elastic force of a material body, an event allegedly involving forces, is conceived from the kinematic standpoint of a factor required to *restore* perfectly elastic collisions—and hence satisfy his momentum conservation law. See Newton, 1962b, ibid., 25.

[12] Snelders (1980, 120). Although it is beyond the scope of this paper to cover this issue in depth, see also, Westman (1980) and Shapiro (1980) for more details of Huygens' physics and its relation to Cartesianism.

[13] See, e.g., Friedman (1983, chapters 2 and 3).

[14] Sklar (1974, 200). Mach's theory is presented in (1942, 280-286).

[15] See Earman (1989, chap. 4), for an analysis of the various attempts by relationalists, including Huygens, to explicate circular motion and its dynamic effects; and chapter 6, for an examination of the viability of (R2), to be discussed below. Earman gives a nice exposition of how Huygens, at least in his later years, tried to preserve (R1) and still account for rotation (see endnote 2). Furthermore, there is the related problem of whether (R1) is implicitly contained in (R2), but there does not appear to be any overt reason for accepting this entailment relationship.

[16] Mach seems to suggest that the Newtonian predictions are simply wrong; i.e., that the rotation of the stars relative to the fixed bucket would likewise result in the centrifugal effects being experienced only by the latter. (see, Mach 1942, 283-284) However, if Mach is correct, so that local inertial effects depend on distant massive bodies, then this fact should be *in principle* verifiable through available experimental or observational evidence (possibly by studying large stellar bodies; see, Sklar 1974, 201)—but, Mach left no hints as to how his theory could be verified by anything less than rotating the entire set of fixed stars.

[17] Westfall (1971, 156-158); Gabbey (1980, 178-181). Huygens' treatment of centrifugal force can be expressed in the modern formulation, mv^2/r, where r is the radius of the circle (Westfall 1971, 170), although Huygens did not carry his discovery that crucial step further, as Newton did, and postulate the force of gravity required to offset this force—the influence of, or devotion to, the Cartesian vortex was crucial , here—see, e.g., Hall (1976).

CHAPTER 9

CONSTRUCTING A CARTESIAN DYNAMICS WITH "FIXED" REFERENCE FRAMES: THE "KINEMATICS OF MECHANISMS" THEORY

Continuing our analysis of the Cartesian response to the *De gravitatione* argument, this chapter will investigate the possibility of avoiding the consequences of Newton's allegations by undermining one of the key premises in its construction: that a matter-filled universe cannot allow unchanging spatial positions. In short, if some sort of fixed reference point can be located in the Cartesian plenum, then a method of tracking a moving body's velocity over time can be established (thus dispelling Newton's anti-relationalist worries). Unlike the procedure adopted in chapters 6 and 8, where we employed the dynamics-based "rest" force or Huygens' notion of a center-of-mass frame to reconstruct a Cartesian dynamics founded upon the collision rules, that is, by using these methods to pick out the privileged frames needed to conserve quantity of motion, this chapter will not appeal to Descartes' specific predictions on the outcomes of bodily collision in order to thwart Newton's argument. Our exclusive concern will be to develop a theory of space and time that will allow a Cartesian to meaningfully conserve quantity of motion, and which does not attempt to utilize any of the collision rules as the basis of this reconstruction. Without the need to maintain the precise predictions of the collision rules (as Huygens' had attempted), it will no longer be necessary to adopt methods that only temporarily preserve quantity of motion from the perspective of a local collision frame. On the formulation of Cartesian dynamics presented in this chapter, we will attempt to posit reference frames that conserve Descartes' quantity of motion for extended regions of the plenum and for extended temporal periods.

With respect to this undertaking, many of the lessons to be gained from studying the modern theory of connected gears, known as the "kinematics of mechanisms," will demonstrate possible methods of locating fixed landmarks in Descartes' universe. In fact, many of the details of the vortex theory proposed in the *Principles of Philosophy* suggest a familiarity with the types of problems normally encountered in constructing series of connected gears. Both theories, for example, accept a view of the "interconnectedness" of all universal motions, and of an apparent lack of "true" collisions (as depicted in the collision rules), that is strikingly similar. As Mark Wilson has suggested (1993), a Cartesian spacetime patterned on the theory of gears may possess the sort of "fixed" landmarks seemingly required by Descartes' physics—a possibility that we shall thoroughly explore.

Admittedly, the Cartesian theory we will develop owes much to recent mathematical and mechanical concepts, which may elicit a number of objections: such as, "How can the modern theory of gears reveal anything about Descartes' seventeenth century theory of motion?", or, "Why bother to investigate Cartesian physics since everyone knows that is obviously wrong?" Although it is certainly true that the forgoing analysis will not deal with Descartes' science in an exclusively seventeenth century setting, the sheer historical importance and, it will be argued, the potential fruitfulness of the basic concepts of Descartes' physics demand a thorough examination—and, if one is to conduct such an investigation, the use of modern conceptual devices can be a particularly effective means of revealing the fundamental strengths and weaknesses of the Cartesian theory. Our use of the kinematics of mechanisms analogy will seem doubly justified, moreover, since Descartes' treatment of the configuration and operation of vortices closely resembles many aspects of the modern theory of connected gears.

Likewise, as noted above, the "Newtonian" world view, broadly conceived, has greatly benefited from the adoption of new mathematical techniques over the preceding three hundred years. The basic laws of motion, the gravitational law, and a (usually weakened) commitment to absolute space have formed the conceptual core of numerous attempts either to consolidate the theoretical advances of a new age, or to resolve specific problems that Newton could not, or did not try to, answer. As an instance of the former, Newton certainly did not foresee the "Full-Newtonian" and "Neo-Newtonian" spacetime incorporations of the revolutionary consequences of Einstein's theories (see chapter 2);[1] nor could he have clearly anticipated, as an instance of the latter, the development of the analytical mathematical techniques, by Euler, Lagrange, the Bernoullis, etc., which eventually allowed the "Newtonian" theory to eclipse the Cartesian vortex theory. Consequently, not only can the employment of recent mathematical methods in studying Descartes' theory provide possible insights into the formulation of modern relational spacetime theories, even those not specifically based on Descartes' ideas, but the use of such techniques may even allow a certain "historical revenge" in that a Cartesian can finally employ the advantages that the "Newtonians" have taken for granted over the last two centuries.

After a presentation of the Cartesian vortex hypothesis and Newton's main argument against it (section 9.1). In the second section, we will compare the details of the kinematics of mechanisms theory with several features of Descartes' vortex hypothesis, thereby preparing the way, in section 3, for a general examination of the viability of constructing a Cartesian theory of space and motion based on the insights gained from the theory of machine parts. In this last section, as well as the conclusion, we will asses the overall success of a suitably altered Cartesian spacetime in resolving two purported internal inconsistencies often raised against Descartes' natural philosophy: first (1), the alleged inconsistency between Descartes' rules of bodily collision and his professed relational view of motion (which, at this point in our investigation, should be a problem that is quite familiar to the reader); and, second (2), Newton's contention, in his early manuscript, *De gravitatione,* that the Cartesian plenum (i.e., a universe completely filled with matter) can not provide the "fixed" spatial locations, or frames of reference, necessary to make sense of Descartes' brand of inertial motion (rectilinear uniform speed). Contrary to the popular verdict, this essay will strive to show that Descartes' handling of space and motion is more consistent than has been generally acknowledged; and, in particular, that a "general" Cartesian theory (based on his fundamental laws and concept of relational motion) has the potential to provide a significant and highly effective response to these two problems if examined with the degree of latitude often accorded Newton's essential hypotheses.

9.1. The Cartesian Vortex and Newton's *De gravitatione* Argument

Developing a consistent dynamical theory for a Cartesian plenum is a rather formidable task. As discussed in Part II, Descartes' hypotheses on the interactions of material bodies generate an enormous range of difficulties when set within a plenum environment. No less problematic, however, are the attempts to construe how bodily motion in a plenum is actually conceived. That is, given the facts of a matter filled universe, what conceptual mechanisms are involved in accurately describing a body's displacement and velocity across time? At this point, Newton's *De gravitatione* argument, first discussed in chapter 1, can be of great assistance in bringing these issues to light. Our analysis of Newton's argument has thus far tended to focus on those more general features of Descartes' theory that are common to most, if not all, formulations of spacetime relationalism. Yet, Newton criticized at length many of the details intrinsic to Descartes' particular union of Aristotelian relationalism and a vortex theory of planetary motion. These details will prove of great importance in determining the viability of the Cartesian program. Consequently, an examination of the pertinent aspects of Descartes' vortex theory and Newton's critique is required.

Due in part to a conviction that a plenum universe could utilize only relational motion, Descartes was ultimately convinced that a treatise on natural philosophy could incorporate the "new" mechanical ideas (see chapter 3 and 7) without running afoul of Church censorship. Essentially, since one of the Inquisition's principal objections to Galilean science concerned the heretical pronouncement that the earth moves, Descartes hoped to avoid this objection by placing the earth within a vortex of secondary element matter circling the sun, and demanding that the earth not alter its position relative to the containing surface of the surrounding material particles (Pr III 24-31). Through this ingenious bit of reasoning, Descartes could then claim that the earth does not move—via his Aristotelian definition of place and motion—and yet maintain the Copernican hypothesis that the earth orbits the sun. "The Earth, properly speaking [i.e., according to the Aristotelian theory of place], is not moved, nor are any of the Planets; although they are carried along by the heaven." (Pr III 28) Overall, the Cartesian universe operates as a network or series of separate interlocking vortices, with each vortex housing an individual planetary system or celestial body. In our solar system, for example, the globule matter within the vortex has formed itself into a set of stratified bands, each lodging a planet, that circle the sun at varying speeds.[2]

When drafting his *De gravitatione* argument, Newton was fully cognizant of the specific features of the Cartesian vortex hypothesis. His primary criticism of Descartes' relational theory of space and time, as noted in chapter 1, concerns its inability to supply the absolute spatial positions deemed necessary to explicate velocity. Without a notion of "same spatial position over time," Newton believed that Descartes' universe could not coherently define velocity (or speed), since there would exist no means of comparing a body's change in position over time. In demonstrating the relationalist's lack of such "absolute" places, Newton appealed to details of Descartes' vortex theory:

> For example, if the place of the planet Jupiter a year ago be sought, by what reason, I ask, can the Cartesian philosopher define it? Not by the positions of the particles of the fluid matter, for the positions of these particles have greatly changed since a year ago. Nor can he define it by the positions of the Sun and the fixed stars. For the unequal influx of subtle matter through the poles of the vortices towards the central stars (Part III, Art. 104), the undulation (Art. 114), inflation (Art. 111) and absorption of the vortices, and other more true causes, such as the rotation of the Sun and stars around their own centres, the generation of spots, and the passage of comets through the heavens, change both the magnitude and positions of the stars so much that perhaps they are only adequate to designate the place sought with an error of several miles; and still less can the place be accurately defined and determined by their help, as a Geometer would require. . . . And so, reasoning as in the question of Jupiter's position a year ago, it is clear that if one follows Cartesian doctrine, not even God himself could define the past position of any moving body accurately and geometrically now that a fresh state of things prevails, since in fact, due to the changed positions of the bodies, the place does not exist in nature any longer (Newton 1962a, 129-130).

As Newton correctly points out, the Cartesian vortex is a system whose constituent bodies and particles are in a constant state of flux. Leaving aside the persistent flow of the first elements of matter (deemed "subtle" matter) between the secondary globules that surround the planets, the capacity of the vortices to radically alter their relative dispositions and internal configuration of particles nicely demonstrates the variable nature of the plenum. In Pr III, Articles 115 and 116 of the *Principles,* for example, Descartes maintains that a vortex may undergo a "shrinking" or (in the modern jargon) collapsing stage, a process which relinquishes the matter of the vortex to its adjacent neighbors. Descartes essentially drafted this complex hypothesis in an effort to reconcile his vortex theory with both the origins of comets and the observed variations in star luminosity: "It can also happen that an entire vortex that contains some such star [at the center of the vortex] is absorbed by the other surrounding vortices and that its star, snatched into one of these vortices, becomes a Planet or a Comet (Pr III 115)." Hence, on Descartes own admission, it would appear that the fixed spatial locations necessary for Newton's determinate trajectories can not be sustained. If one were to attempt to utilize a body in the plenum, or the contiguous particles surrounding a body, as a relational means of securing the fixed spatial locations needed to determine velocity, via the (R2) or (R1*) classification introduced in chapter 6, the flux of the plenum would consistently conspire to thwart this process by dislocating and disintegrating the relational reference frames. Due to the fluctuating nature of the plenum, any planet or star (or body) runs the risk of a forced separation from its native vortex and its contiguous secondary elements of matter, thus spoiling the hopes for an effective relational coordinate system. (This problem with Descartes' theory is labeled (2) above.)

As revealed in the above quotation, Newton's argument contains a strong epistemological component; an aspect of his reasoning that was not specifically disclosed in our discussion from chapter 1. That is, through his elaborate analysis of the variable nature of the plenum, Newton seems intent on demonstrating that the Cartesian theory of place and motion can only provide, at best, a very inaccurate approximation of the positions, and hence velocities (speeds), of bodies over time: e.g., "[the positions of the stars] are only adequate to designate the place [of Jupiter] with an error of several miles." Since the stars are likely to alter their relative positions due to the ceaseless flux of the plenum, they cannot furnish reliable estimations of place and motion. In fact, it would seem that Newton is combining two separate arguments against the Cartesian theory at this particular point in the *De gravitatione* presentation: (1) the epistemological criticism just noted, which centers upon the plenum's inability to secure accurate measurements of a body's place and motion, and (2) the ontological problem, much discussed in chapter 1, that "due to the changed positions of the [moving] bodies, the place does not exist in nature any longer." In other words, when a given body moves, its place no longer exists, since "place" is defined as the common surface between the contained and containing surface—a surface which is irrevocably lost once the displaced body takes on a new set of contiguous neighbors. (See the passage quoted in chapter 1: both the above quotation and the passage provided in chapter 1 appear in the same context, and page, in the *De gravitatione*.) This form of reasoning, in my opinion, is best interpreted as an ontological criticism of Descartes' theory; for it claims that the very *concept or meaning* of velocity (speed) is not definable given the Aristotelian/Cartesian doctrine of place. Although the main emphasis of Newton's argument in the *De gravitatione* involves this ontological critique, as duly observed in chapter 1, the epistemological aspect of his contention with Descartes should not be overlooked.

Of course, both the epistemological and ontological aspects of Newton's argument against Descartes' theory of space and time rely upon his notion of an "absolute spatial position:" regardless of whether we are attempting to measure or to define velocity, it is necessary that our spacetime be equipped with absolute spatial positions—a concept that we have reason to reject provided the analysis of chapter I. Nevertheless, Newton's insight, that a plenum without absolute space constitutes an environment hostile to the notion of velocity, acceleration, etc., will assume great importance in the remainder of our investigation, especially when translated into the context of our search for an (R2) or (R1*) reference frame. In examining the details of a Cartesian spacetime modeled on the theory of gears, we will return to Newton's views on the plenum and motion.

9.2. The Kinematics of Mechanisms and Cartesian Spacetime.

In this section, we will begin to explore the possibility of constructing a Cartesian spacetime that can overcome problems (1) and (2), from introductory section to this chapter, while conserving a relational quantity of motion, or size times speed, by means of fixed reference frames. This spacetime will also need to closely adhere to the spirit of Cartesian physics by taking into account the kind of harmonious, interconnected motions typical of a Descartes' plenum. In order to examine these issues more closely, it will be fruitful to correlate the Cartesian program with a more recent mechanical theory that investigates many aspects of equivalent problems. Entitled the "kinematics of mechanisms" (KM), this branch of physics analyzes systems of rigid mechanical linkages, such as an array of connected gears. On the whole, many of the worries that would motivate an engineer in constructing an elaborate series of gears relate directly to the obstacles encountered in attempting to comprehend motion in the Cartesian plenum, especially with respect to vortex motion.

9.2.1. Vortices and Gears Compared.

The theory of machine parts and Cartesian physics are similar in many ways. In an elaborate set up of gears, for example, the movement of one cogwheel entails a determinate motion of all the other cogwheels connected to the system. This parallels a similar situation confronted in the Cartesian universe where the displacement of one particle inevitably results in the vast (circular) movements of others: "It has been shown . . . that all places are full of bodies . . . . From this it follows that no body can move except in a complete circle of matter or ring of bodies which all move at the same time (Pr II 33—see, also, chapter 5)." Circular motion is necessary for Descartes because there exist no empty spaces for a moving object to occupy. If the motion were not circular, the movement of a single body in his "indefinitely" large universe would result in an equally indefinite material displacement, an outcome that may violate the Cartesian conservation law. Consequently, on Descartes' own admission, the predominate interactions among the material constituents of a plenum, as well as for an assembly of gears, would seem to be some form of interlocking uniform motion, and not the direct impact as spelled out in his collision rules.

Descartes' concept of an "inclination" towards motion fits naturally into this comparative project, moreover. In attempting to achieve the goal of a kinematically-oriented description of material body interactions, Descartes located a body's "tendency" or "inclination" to move, which can be roughly conceived as its "force" of motion (measured by quantity of motion—Pr II 43, see chapter 3), at the level of individual instants; while conceiving motion as a process that only occurs over a finite temporal interval. "No movement is accomplished in an instant; yet it is obvious that every moving body, at any given moment in the course of its movement, is inclined to continue that movement in some direction in a straight line (Pr II 39)." The Cartesian objective, in simplest terms, is to analyze dynamic interactions, which seem to involve bodily forces, while confined to the language of the relative motions of particles, a purely kinematic approach that is directly analogous to the "interactions" of connected gears (see, Garber 1992, 218-221, for an extended discussion of Cartesian "tendency"). Descartes illustrates the underlying affinity or kinship of these two programs in a very revealing passage from the *Principles* concerning the "strivings" of secondary element matter:

> When I say that these little globules strive, or have some inclinations, to recede from the centers around which they revolve, I do not intend that there be attributed to them any thought from which this striving might derive; I mean only that they are so situated, and so disposed to move, that they will in fact recede if they are not restrained by any other cause. (Fr Pr III 56)

If "not restrained by any other cause", bodies will thus move because "they are so situated, and so disposed." When we conjoin this hypothesis with his belief that inclinations or strivings are instantaneous, the following picture of Cartesian "inclinations" begins to emerge: on each time slice, the configuration of all plenum bodies—their relative disposition and situation—determines how a single body in the system *can* move; that is, how the arrangement of the whole affect or inhibits the motion of a single body. Much like machine parts, the possibilities for movement of a single body are governed by the interconnections of all material objects, since a given body can only move if the resulting displacement of bodies is harmonious and does not lock-up the system (which explains the reason for mass circular motions noted above). Or, to put it differently, the configuration of all bodies determines how the motion of one body instantaneously affects all the others (as with connected gears). Finally, if one envisions a succession of several instants, the inclination of a body at each separate instant gives rise to its speed, and thus the measurement of its quantity of motion; since, as Descartes states, speed is a higher-level phenomenon that is manifest over a span of time.

Gears and plenums likewise must be designed so that the motion of their constituent parts are compatible and harmonious, and will not lock or jam. A "lockup" can occur in machine parts when the motion of single gear is prevented through its connection to two oppositely rotating cogwheels. For the Cartesians, this complication would translate into a collision, or "blending," of particles from several divergently rotating vortices. Overall, Descartes was well aware of the need for an effective and harmonious positioning of neighboring vortices: "No matter how these individual vortices were moved in the beginning, they must now be arranged in harmony with one another so that each one is carried along in the direction in which the movements of all the remaining surrounding ones least oppose it (Pr III 65)." In Pr III of the *Principles,* Descartes presents a number of hypotheses on the mutual ordering of vortices that reveal a deep understanding of the problems that can beset arranging mechanical systems. Even though many of these constraints on the ordering of vortices are intended to satisfy his hypothesis on the flow of subtle matter, the above quotation is clearly aimed at forestalling the collision or interference of adjacent vortex rotations. Within these Articles, Descartes describes in painful detail the configuration that is necessary to prevent the "opposition" or clashing of the rotational motion of four neighboring vortices:

> The laws of nature are such that the movement of each body is easily turned aside by encounter with another body. Accordingly, if we suppose that the first vortex, the center of which is S, is rotated from A through E toward I, the other vortex near to it, the center of which is F, must be rotated from A through E toward V if no other nearby vortices prevent this; for thus are their movements most compatible. And in the same way, the third vortex, which has its center, not on the plane SAFE, but above it (forming a triangle with the centers S and F), and which is joined to the other two vortices AEI and AEV on the line AE, must be rotated from A through E upward. (Pr III 65—see Figures 25 and 26)

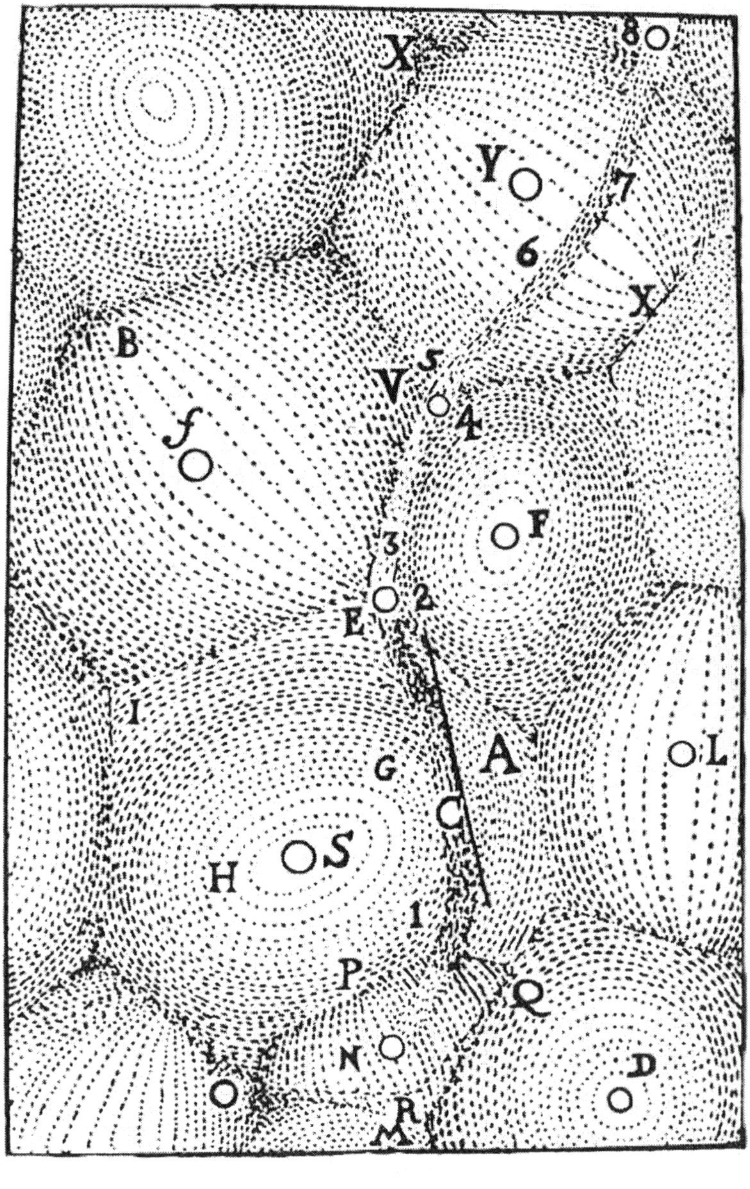

Figure 25. Descartes depiction of the plenum vortices in the *Principles,* Pr III 65.

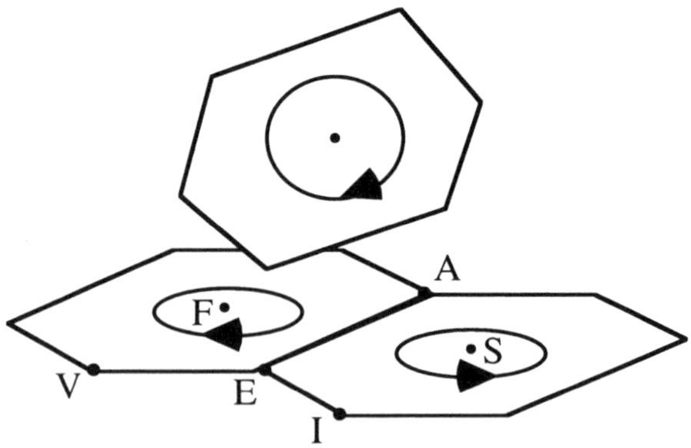

Figure 26. This is a simplified illustration of the harmonious configuration of Descartes' vortices in the *Principles,* Pr III 65 (Plate VI). The third vortex, which is suppressed in Descartes' original illustration, lies above the plane of the other two.

9.2.2. Kinematics of Mechanisms and Cartesian Spacetime.

In devising a Cartesian spacetime[3] according to KM, certain requirements need to be met. (The technical details of this section owe much to the discussion in Wilson 1993, 216-218.) First, a time function must be established that partitions the events in spacetime into simultaneity classes. This is usually depicted as the carving of spacetime into a series of "time slices" or spatial planes, with each slice representing all the existing material bodies at a particular temporal instant. Second, the spatial geometry on each slice must be three-dimensional and, probably (but not essentially), Euclidean[4]; and a time metric is imposed on the spacetime to uniquely order the time-slices. So far, these conditions can also be found in Newtonian spacetime, but not enough structure has been added to make it obviously "absolutist." Since we want to insure that particles can be tracked through time while preserving the topology of their local connections, the spacetime of our connected gears requires a device or function that will identify the same material particles or points between time slices. This is accomplished by defining a map Ψ from the material points on a reference set of machine parts (or bodies) to the same points on each slice. The mapping Ψ ensures that our mechanical gears remain rigid over time by maintaining the same distance relations among their various parts and material points (see Figure 27). In addition, all the machine parts located on a time slice must be interconnected via some mechanical process that prevents the slipping of gears at their contact points (i.e., the point where two meshed gears touch). Suppose, for instance, that two material points, p and q, that are situated on separate mechanical gears are in contact at a time t_0. Next, divide the relative arc length displacement $\overline{d}_p$ of p away from the previous contact point (on t_0) by the change in time Δt (i.e., $t_1 - t_0$) to obtain the value $\bar{d}_p/\Delta t$. One can avert the slipping of gears, consequently, by demanding that $\bar{d}_p/\Delta t$ varies smoothly and equals $\bar{d}_q/\Delta t$ as the limit of Δt approaches zero. (The relative displacement of points on rotating gears can also be determined by fixing three points. See, Zimmerman 1962, 29.)

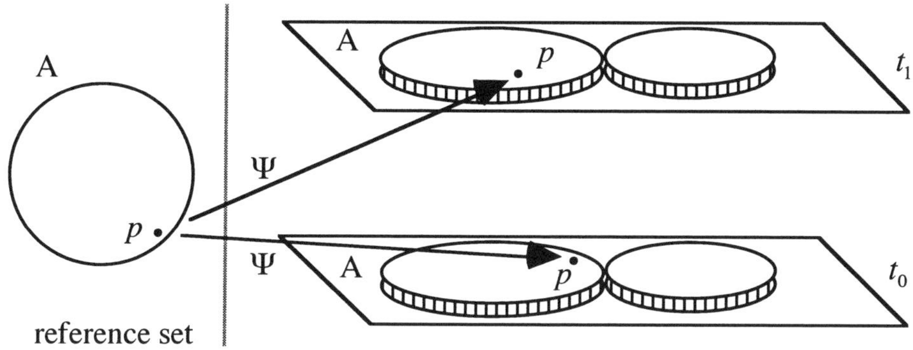

Figure 27.

Another important component of KM is the notion of a "fixed space" or landmark between temporal slices. Ideally, if one wishes to employ a *relative* velocity function in our spacetime, it will be necessary to establish temporally fixed reference frames so that the map Ψ can assign a relative displacement, and hence velocity, to all material particles. This can be accomplished in a number of ways: one can simply tie the reference frame to a material point picked out on each time slice by the mapping Ψ; or one can choose an enduring geometric feature of the overall configuration of gears as the preferred reference point.[5] On the latter procedure, a contact point between two gears would naturally serve the role of a fixed space in our theory, since the only permanent locations on the contact surface of moving machine parts are the places where they touch. Hence, provided Ψ and the mapping Φ of a unique contact point across temporal slices, the velocity of a point p (on a gear A) is easily obtained by measuring the difference in relative arc length displacements $\bar{d}_p$ of p between slices (see Figure 28—the mapping of p from the reference set Ψ has been suppressed in this illustration). It should be noted, furthermore, that there is no limitation on the temporal duration of the Ψ and Φ mappings (although $\bar{d}_p/\Delta t$ is instantaneously defined, of course, as the limit of $\bar{d}_p/\Delta t$ as Δt approaches zero). The temporal extent of these mappings depends on the stability of the gear system. However, in Section 9.3, we will explore the possibility that the intrinsic instability of the Cartesian plenum may confine these mappings to a very brief temporal duration (perhaps infinitesimal, just as long as it can allow the limit of $\bar{d}_p / \Delta t$ to be established).

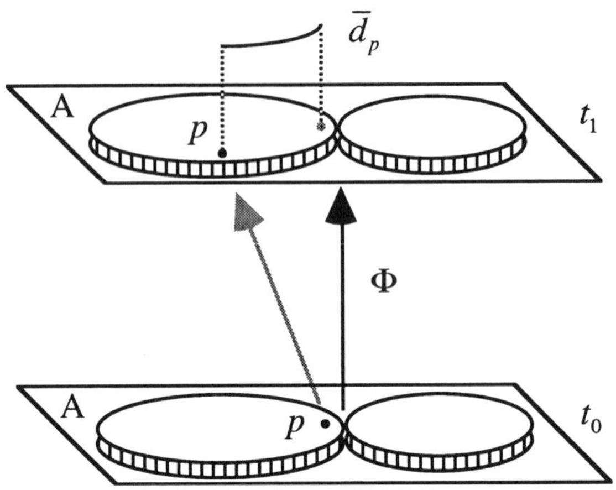

Figure 28.

In many respects, the details of the KM program correlate nicely with our earlier characterization of Descartes' plenum theory. For instance, via the natural laws (as above), Cartesian spacetime invokes a notion of rectilinear uniform speed, or velocity,[6] that necessitates a comparison of "information" across time slices, a concept that also features in the theory of gears. More specifically, the motion of both a Cartesian body and a point on a gear can only be determined over a series of successive temporal slices. One must examine the difference in a body's displacement between slices (relative to a reference frame)—the "information" is simply the displacement on each slice—in order to attribute this property to the body on a single time slice. Of course, KM, like all modern dynamic theories, has a perfectly meaningful concept of instantaneous velocity defined at an instant; namely, the derivative of the position function. But calculating the derivative of a particle trajectory requires a span of time: as defined above, it is the limit of the function as the change in time approaches zero. The motion of a Cartesian body, if not incorporating infinitesimal quantities of this exact sort, is nevertheless characterized in a similar way; i.e., as requiring some means of *transmitting information across temporal slices* on a body's displacement (although Descartes does not employ these sophisticated mathematical techniques). Moreover, even if Descartes' understanding of motion cannot be isolated from duration or a succession of time slices ("no movement is accomplished in an instant," (Pr II 39)), his conservation law for the quantity of motion does appear to implicate instantaneous processes or magnitudes (e.g., his account of "strivings", or "the force to continue in motion", Pr III 121-122, as above). Thus, Descartes' theory is much closer to the instantaneous magnitudes of the modern formalism than has been often acknowledged.[7]

If Cartesian speed can only subsist over a series of time slices, many tenets of KM may look appealing to a relationalist engaged in reworking Descartes' laws of motion. Foremost among these properties of the theory of gears is the use of the landmark mapping Φ to ground the measurements of relative velocity, since these landmarks would essentially guarantee the determination of relational motion as mandated in chapter 1. If some form of analogous *enduring* landmark can be established in the Cartesian plenum, a relationalist can provide a coherent system for determining relative speed according to the tenets of (R2) or (R1*) by simply constructing a relatively "fixed" reference frame for an extensive region of Descartes' plenum, possibly even the entire cosmos. From these landmarks, the *individual* motions of all material bodies will be measured, thus securing a means of resolving relational contradiction between, say, the fourth and fifth collision rules—this problem we labeled (1) above in the introductory section. As explained in chapter 3, the non-relational undercurrent in Descartes' natural laws is most evident in the case of collision rules four and five: briefly, in the fourth rule, a large object remains at rest during impact with a smaller moving body, and simply deflects the smaller body back along its path (Pr II 49); whereas in the fifth rule, a large body will move a smaller stationary object, "transferring to [the smaller body] as much of its motion as would permit the two to travel subsequently at the same speed (Pr II 50)." Conceived in a relational manner, however, rules four and five constitute the same type of collision, for they both involve the interaction of a small and large body with the same relative motion (or speed difference) between them. Since they represent identical phenomena from the relational standpoint, a consistent relational theory demands an identical outcome; which Descartes does not provide. Nevertheless, given the KM landmark procedure outlines above, one can give a meaningful measurement of the individual states of motion of the impacting bodies, and thus the types of collisions depicted in rules four and five can now be consistently regarded as relationally different or unique. In rule four, the larger body really is at rest and the smaller body really is in motion (relative to the reference frame, of course), and visa-versa for rule five.

However, before the Cartesians can endorse the landmark mapping Φ as a means of resolving the inconsistency in Descartes' handling of the collision rules, problem (1), they will need to refute Newton's allegations concerning plenum instability, our problem (2). As translated into the language of KM, Newton's argument contends that a landmark or Φ based conception of linear uniform speed ultimately fails due to the lack of a constant Φ in Descartes' universe. Unlike our series of connected gears, there exist no immutable landmarks or contact points in the Cartesian plenum to ground the computations of these motions. Consequently, if one accepts the view that Cartesian spacetime resembles a series of connected gears, Newton would have to insist that nature persistently conspires to detach the connections among the machine parts, thus dispelling the contact points. In fact, the constant flux of the primary particles in Descartes' universe nicely demonstrates the mutability of the connections among the "gears" in our Cartesian kinematics of mechanisms analogy: just as the particles in the plenum consistently change their relative position, a similar process must transpire as regards the gears in our Cartesian machine-part universe. That is, the gears (particles) will partake in a continuous shifting of their relative positions and a changing of their mutual connections, meshing and unmeshing with a host of different particles (gears).

Alternatively, one can interpret Newton's argument in this context as a straightforward denial of the smoothness of the connections among the particles (gears) required for $\bar{d}_p/\Delta t$. Likewise, because there are no completely impenetrable or indestructible bodies in Descartes' plenum—all material elements can shatter under sufficient impact (see, e.g., Pr II 20, and II 48)—rigidity cannot be guaranteed. As a direct consequence, either the slipping of the particles (gears), or their lack of rigidity (i.e., the Ψ mapping cannot maintain a fixed separation among the parts), can be viewed as primarily responsible for the lack of a workable landmark-based measure of motion. Regardless of the particular cause of this instability, a world without permanent contact points is a world that still requires a permanent notion of uniform speed (velocity). As presented in the *De gravitatione,* Newton's "Jupiter" example was intentionally designed to demonstrate this very point: although the Cartesian plenum exhibits no enduring landmarks, it is clearly the case that the trajectory of Jupiter is determinable.

9.3. Developing a Kinematics of Mechanisms Cartesian Spacetime

9.3.1. Locating Φ-Landmarks in the Cartesian Plenum.

But not all hope is lost for the Cartesian, for we have yet to examine the available options for locating the elusive fixed landmarks in Descartes' plenum that can evade our problem (2), and thus obtain a means of resolving problem (1). In fact, as previously argued, a close inspection of the Cartesian vortex theory can reveal many hidden facets of Descartes' understanding of mechanical systems; an awareness of the complexity of vortex ordering that may suggest possible methods of devising Φ-landmarks.

Among the basic operating principles of the vortex, the movement of the first particles of matter (or subtle matter) figures prominently. Briefly, Descartes reckons that a significant amount of subtle matter perpetually flows between adjacent vortices: as the matter travels out of the equator of one vortex, it passes into the poles of its neighbor. This hypothesis is an integral component in his story of vortex "collapse" (Pr III 115-120). Under normal conditions, particles of subtle matter flow from the poles into the center of the vortex (i.e., the sun); then, due to centrifugal force, the particles "press out" against the surrounding secondary globules as they begin their advance towards the equator (ibid., 120-121). However, since the adjacent vortices also possess the same tendency to swell or increase in size, centrifugal force prevents the encroachment of neighboring vortices by setting up a balance of mutual expansion forces. On occasion, a debilitating condition of the sun (identified as sun spots) may conspire to prevent the incoming flow of first element matter from the poles. Gradually, as all the first matter is expelled at the equator, the sun can no longer press against the secondary globules, and the vortex is engulfed by its expanding neighbors.

For our purposes, an analysis of these "intervortex" relationships is important because it may provide a suitable candidate for an invariant or unchanging contact point. Although the possibility of vortex collapse effectively dashes any hope of locating a permanent landmark within a vortex, it does not automatically dismiss the potential existence of such landmarks *between vortices*. In particular, since Descartes envisions vortex collapse as a gradual expansion process, the contact points will be continuously maintained between the remaining adjacent vortices. These mutual connections may gradually shift or alter position as the rotating masses increase in size, but the enormous forces that lock the vortices together can, it is hoped, ensure that the contacts are not dissolved by separation, and that any displacement or dislocation of these points will *generally* occur smoothly without slipping (as required to supply the measurements of velocity). In contrast to a more random, chaotic configuration of vortices, the enormous pressure exerted between the interlocked vortices will, on this reading of Descartes' project, greatly decrease the potential for the slipping of particles (gears). Of course, this strategy cannot completely guarantee that only smooth connections will prevail among Descartes' vortices; and, hence, it can only diminish or mitigate the force of the Newtonian argument. Nevertheless, a Cartesian can go a long way towards resolving Newton's problem by employing these intervortex contacts as the basis for a mapping Φ. This mapping would effectively equip his spacetime with a quasi-fixed reference frame or class of such frames—the intervortex contact points—for the (R2) or (R1*) relational determination of bodily motions within each vortex (as noted in section 9.1). Finally, it should be noted that the idea of a "contact point" in a Cartesian plenum is somewhat of a misnomer, since only the three primary elements of matter would retain the necessary rigidity to provide anything remotely close to a point-like connection between surfaces. Since Descartes views most, if not all, macroscopic bodies as "elastic" (due to the presence of bodily pores which can contract under pressure; see, Pr IV 132), the contact between vortices would more resemble an extended two-dimensional surface or band, rather than a point.[8]

Besides the ever present threat of non-smooth connections, Newtonians may dispute the legitimacy of this stratagem on other grounds, however. They would most likely insist that Descartes never envisioned the contact points among vortices as a prerequisite structure for his concept of speed (velocity). He defined motion, you will recall, as the transfer of contiguous bodies *within* a vortex; a process, moreover, that would not seem to necessitate the existence of any intervortex relationships. Yet, as argued above, modern "Newtonians" (of either the Full-, or Neo-, variety) are not in a favorable position to criticize others for taking conceptual liberties with historical spacetime theories, since the product of their efforts are a far cry from Newton's account of these issues. For example, would a reemergent Newton accept the Neo-Newtonian metric tensor, $\mathbf{g}(\mathbf{V},\mathbf{V})$, and covariant derivative, $\nabla_{\mathbf{V}}\mathbf{V}$ (or $d^2x^\alpha/d\lambda^2 + \Gamma^\alpha_{\mu\nu}(dx^\mu/d\lambda)(dx^\nu/d\lambda)$ in a coordinate form), as an updated reflection of his conception of absolute space and time (which was as embedded in theological concerns as much as it was in physical problems)? In essence, there is much to be learned about the structure and limitations of *both* the Newtonian and Cartesian theories when set within the modern mathematical formalism, hence the modern Cartesian should be allowed equal access to these techniques.

As a second difficulty with the intervortex contact points, even if one permits a Cartesian to exploit this method, is securing a consistent basis for velocity in a potential world consisting of only one vortex (or none)? Inasmuch as contact points require at least two vortices, one could not locate a coherent notion of velocity in such a universe, while any Cartesian efforts to exclude this scenario from the domain of possible evolutionary states of the plenum would appear an unjustified and ad hoc restriction. Interestingly, in the *Principia,* where the main assault on the Cartesian vortex theory appears (in Pr II), Newton contends that the inherent instability of the stratified layers in a planetary vortex, such as our solar system, would inevitably lead to their "blending" and certain destruction (by possibly forming one large undifferentiated universal vortex?—Newton, 1962b, 391).

In response to these troubling possibilities, a Cartesian may fall back on the text of the *Principles* and point out that these scenarios are simply ruled out as possible states of Descartes' universe; or, in other words, since Descartes did not discuss it, it must not be possible in his plenum. Despite the unappealing tone of this form of response, Newton's allegation of vortex instability does force one issue to the forefront of our attempts to locate a Φ-landmark: a Cartesian may decide in the light of these difficulties to abandon the search for a permanent landmark, while concentrating instead on the class of transitory or temporary contact points between vortices. Given the details of Descartes' cosmology, wherein he assumes that there will always exist numerous vortices, temporary contact points (or surfaces) will always exist. Moreover, many such points also exist in a planetary system, since each planet is transported around the sun in its own circling band of secondary particles. Hence, the centrifugal force exerted by these rotating bands will guarantee the availability of an entire class of potential "intravortex" landmarks. The circling bands that comprise a vortex will be locked together by their mutual expansion, thus (hopefully) providing an array of smooth connections (notwithstanding the difficulties mentioned above). Although the fluctuating nature of the plenum dictates that these contact points will endure only temporarily, possibly only infinitesimally, it is equally true that there will always be a landmark available. These landmarks will change over the course of time, and may be of infinitesimal duration, but every time slice will have recourse to at least one Φ-based frame for determining motion. Therefore, adapting Descartes' natural laws to a set of temporary intervortex, or intravortex, landmarks might serve as the Cartesian's last line of defense against the Newtonians.

It should be noted that the kind of temporary landmarks considered in the context of our KM reconstruction of Cartesian dynamics, and their corresponding temporary reference frames, differ from the temporary reference frames developed in chapter 6 and 8. As regards our KM variant, the temporary frames are not conceptually linked to Descartes' collision rules, as they are so linked in the case of Huygens' center-of-mass frames. The KM frames (as noted in the introduction to this chapter), whether constructed for a finitely or infinitely fixed span of time, are conceptually independent of the collision rules and the impact of particular bodies; thus they are best viewed as ranging over a much greater portion of the plenum and over a much longer temporal period. (Nonetheless, as mentioned in section 6.4, a less constrained interpretation of the dynamics-based "rest" force method might allow such frames to be conceived apart from the context of the collision rules.)

A Newtonian may willingly grant the Cartesian this procedure for establishing landmarks, since such reference frames are only temporary and not permanently fixed. Nevertheless, it remains unclear if such methods can successfully provide the foundation for a coherent notion of speed (velocity); that is, if it can resolve the relational contradiction between collision rules four and five, which we defined as falling within the context of problem (1). In the next section, we will examine problem (1) updated to the context of KM, for several important insights into the nature of Cartesian spacetime can be gleaned from such a setting.

9.3.2. Dead Points and the Collision Rules.

There are various ways of interpreting the nature of problems (1) and (2). One method of construing problem (2), for instance, is to regard Newton's argument as pertaining to the lack of a suitable method of comparing information on the status of bodies, particularly, position, between time-slices. Viewed in this fashion, Newton's argument amounts to a demand for a "geometrical background structure," as described in Part I, to ground the measurements of bodily motions. Due to plenum instability, the ties between time slices, Φ, that serve as the means of comparing information on the positions of bodies relative to a Φ-based frame (i.e., the geometrical background structure), are constantly severed; which brings about a corresponding loss of information on the positions, and hence motions of bodies (or "striving", if we consider Descartes' quasi-infinitesimal quantities). Yet, similar "under-determination" difficulties, as we may dub this information-transfer debacle, also arise even if we grant the Cartesian a regular system; that is, even if we ignore Newton's instability argument (problem (2)). To demonstrate this point we will need to return to our analysis of machine parts.

In constructing an array of connected linkages, engineers strive to eliminate what are deemed "dead points" from their chosen configuration. Briefly, a gear reaches a "dead point" when its future motion is not determined by the instantaneous actions or motions of the other gears located on the same time slice. Two options are generally presented at such points: either the linkage can proceed forward or reverse its direction. A "dead point" in the motion of two connected gears is often presented in the scenario depicted in Figure 29 (see, Wilson 1989, 509-512). Assuming the left wheel is driven counter-clockwise, when the linkage reaches the position at time t_0 illustrated in (i), it has two options: It can proceed along the same circular route (ii), or it can reverse its direction (iii) resulting in one of the two depicted scenarios at time t_1. In either case, the configuration of the gears cannot, by itself, determine the unique evolution of the system beyond t_0 (i). Yet, this outcome is not constrained in any way by the current dispositions of the surrounding gears. One must bring forth additional information or methods in order to determine the future course of the machine parts. "In practice, dead points must be avoided or external means provided to carry the mechanism past a dead point." (Zimmerman 1962, 123) Often, the method or means of carrying a gear past a dead point is its inertial motion. The unique evolution of the gear assembly is determined by the inertia retained by the flywheel, an "absolutist property" that prevents the gear from reversing its direction at the point in question.

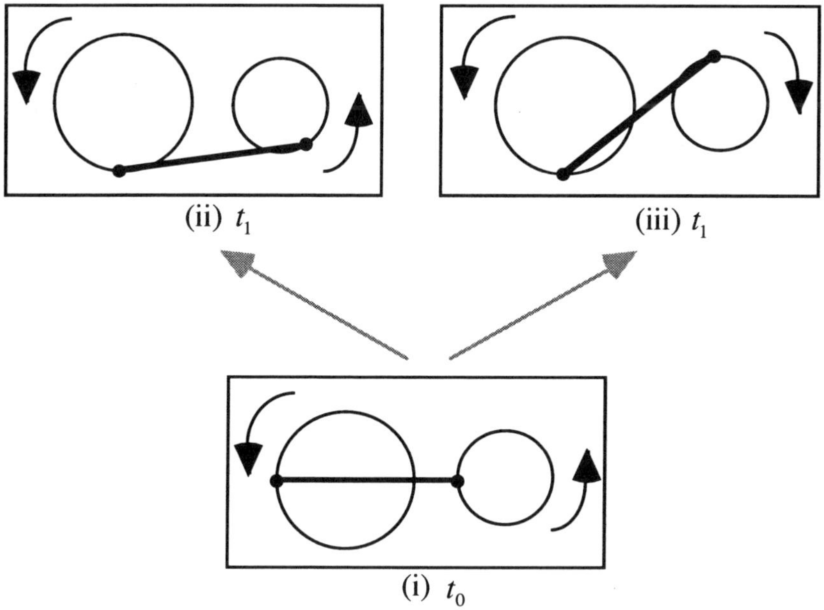

Figure 29.

If the problem of the relational incompatibility of Descartes' collision rules, i.e., problem (1), is viewed in this context, then the phenomenon of dead points bears a striking resemblance to the relationalist puzzle of Descartes' handling of rules four and five. In the gear analogy, the configuration and motions of the gears is exactly the same in cases (ii) and (iii) *before* the dead point is reached, and after, or at, this point their behavior diverges radically. Likewise, in Descartes' plenum, the relational configuration and motions of the bodies in both rules four and five is exactly the same *before* their collisions, but the outcomes of these collisions differ considerably; which, given their identical relational scenarios, should not be the case. Hence, both the mechanical engineer and the Cartesian physicist are faced with an under-determination problem; an under-determination that occurs despite the *stability* of their respective systems (since problem (2) is not in effect). Just as the gear behavior after the dead point is unknowable given the setup, so the behavior of Descartes' colliding bodies is unknowable given his formulation of natural philosophy. The combination of Descartes' collision rules and relational motion has resulted in a corresponding inability to determine what occurs after a specific type of bodily impact (between a smaller and larger body): if we accept the account in the collision rules, then relational motion is violated; and if we adhere to his relationalism, then the outcome of the collisions must be the same (which they are not). Basically, problem (1) amounts to a Cartesian version of

KM dead points; although, of course, Descartes' under-determination is rooted in the contradictory *theoretical* or *conceptual* demands he has made on his physical system, whereas the gear under-determination stems from an unfortunate, but avoidable, material arrangement of the gears.

The mechanical engineer can overcome the dead point difficulty by merely allowing the inertia of the gear, or flywheel, to carry it past the dead point, thus eliminating outcome (iii). In other words, the directed motion of the gears in (i) is preserved in (iii) without the sudden stopping and reversing, deceleration and acceleration, exhibited in (ii). For a Cartesian, the means of overcoming the relational quandary of collision rules four and five, as detailed in section 9.2, is to posit fixed reference frames that can provide the individual components of motion of the colliding bodies (and thus delineate instances of rule four from rule five). If the Φ-based reference frames endure only temporarily, that is, for finite periods of time rather than infinitesimally, the existence of one of these frames throughout the time spanned by the collisions will allow the Cartesian to distinguish instances of the two collision rules, and hence eliminate the sort of theoretical under-determination implicit in Descartes' handling of the cases. On the other hand, an infinitesimal rendition of the Φ mapping will fail to resolve the problem of dead points, most notably if the frame is established at t_0. A frame attached to an infinitesimal Φ mapping at the instant of collision would seem to be incapable (by its very instant-based construction) of tracking the motion of the bodies past their impact, and thereby removing the inconsistency of rules four and five. In fact, a much greater problems lies just underneath the surface, and it is largely due to Descartes' "discontinuous" treatment of the motion of impacting bodies. He argues, AT III 592-593, that bodies do not gradually acquire or lose speed during collision; rather, they achieve their post-impact speeds instantaneously. Needless to say, such a view spells trouble for any attempt to establish our instantaneous $\bar{d}_p/\Delta t$ concept of velocity, since the $\bar{d}_p/\Delta t$ quantity will break down at this point. (This problem also nicely demonstrates the advantages of a principle of continuity, like Leibniz's, that can apparently evade these difficulties.)

The possibility of installing Φ-based reference frames leaves the Cartesian with one more major decision, however: a single privileged Φ reference frame, or set of such frames (presumably at rest relative to each other), must be selected out of the class of all frames. Moreover, since these frames are temporary, as above, and will be in operation only as long as the Φ-landmark endures, a new privileged frame will need to be selected with each demise of the old select landmark. A major difficulty would present itself to the Cartesian if this choice of an a exclusive reference frame were not made; for although the frames can distinguish instances of, say, rule four from rule five, which collisions become categorized as falling under one rule or the other depends on where you position the reference frame in Descartes' plenum. If the frame is placed alongside, or on, the smaller object in rule four, and this frame moves in tandem with the body, from its perspective the body will remain at rest.

This means that it will view the collision with a larger stationary body as a collision with a larger moving one, or, as an instance of rule five (small resting body vs. large moving body), and not rule four (small moving body vs. large resting body). Other frames not attached to, or moving with, the smaller body will, obviously, view the smaller body as at rest, i.e., as an instance of rule five. (A similar situation would arise in our gear analogy if a Φ-landmark were attached to a fixed material point on the edge of the smaller gear in (iii), since its motion backtracks upon reaching the dead point.) Consequently, the relational problem reemerges once again, but this time at the level of the reference frames, rather than with the impacting bodies. Unless the Cartesian is willing to admit conflicting ascriptions of the collision rules to the very same collision—an outcome that would render the application of Descartes' natural laws disastrously relative—they will need to privilege one particular frame, or a certain set of frames among the set of all possible frames, thus guaranteeing a unique or equivalent determination of motion for the duration of that frame.

9.4. Invariant Universal Quantities of Motion

Despite the problems raised by our hypothetical Newtonian in the last section, a Cartesian may appeal to Descartes' conservation law, as exemplified in his claim that God "always maintains in [the world] an equal quantity of motion (Pr II 36)," to furnish a basis for a privileged set of reference frames, and thus ensure a comparing of displacement information across time. If the same magnitude of quantity of motion is preserved at each instant, then all the time slices are bound by a special relationship that induces a form of information transfer. More carefully, given the configuration of interacting particles on any one time slice, the Cartesian conservation law places a constraint on the future evolution of that state, since only those spatial planes whose gears preserve the invariant universal quantity of motion are admissible. Accordingly, the Cartesian has acquired a method for meaningfully comparing the material disposition of the entire plenum across time slices. This form of explanation nicely correlates with the dynamics-based "rest" force approach developed in chapter 6, for we are, once again, allowing the Cartesian conserved quantity of motion to establish, or "bootstrap," the privileged frames for measuring quantity of motion.

To illustrate this point, consider our series of connected gears: at any one time, the motions of the machine parts are governed by a conservation law that, if preserved, demands that the entire system compensate for the alterations in speed of a single gear. Hence, if a machine part were removed or added, the remaining gears would have to change their speeds to preserve the same overall quantity.[9] In a sense, the gears must contain a kind of "memory" of this desired invariant, since they perpetually seek to offset any variations in the status of their fellow parts. Moreover, this conservation law also allows a certain predictive scope, since it will be possible to predict the velocity of any further cogwheels added to the system (based on their size and the overall conserved quantity).

A similar process occurs in the Cartesian plenum. As mentioned, the collapse of a vortex entails the encroachment of its contiguous neighbors, and thus a large scale reconfiguration of plenum inhabitants. Nonetheless, throughout this massive reshuffling of bodies, Descartes' quantity of motion is conserved at each succeeding instant, an outcome akin to the removal of machine parts just described. This insight into the vortex mechanics of the Cartesian plenum provides, moreover, a useful rejoinder to the Newtonian allegations outlined in section 9.2 and 9.3. When the arrangement of the particles situated on a time slice invariably changes, as Newton insists, the particles will necessarily assume a new configuration that *preserves* the total universal quantity of motion (since the conservation law governs the behavior of bodies both across time and at instants of time). For example, suppose a particle (or vortex), a, of size m' and velocity v' were to be removed (in some fashion) from Descartes' plenum, given the invariant universal magnitude of quantity of motion mv, the remaining particles (or vortices) will need to adjust their speeds to compensate for the loss; which, in this case, will require an overall increase of speed, v'', of the remaining bodies, m'', to identically match the lost quantity $m'v'$ ($m''v'' = mv$, see Figure 30). In short, the conservation law regulates the flux of the plenum over time (motions of the particles) by strictly controlling the acceptable arrangements of material particles at each instant (the "tendencies" towards motion).

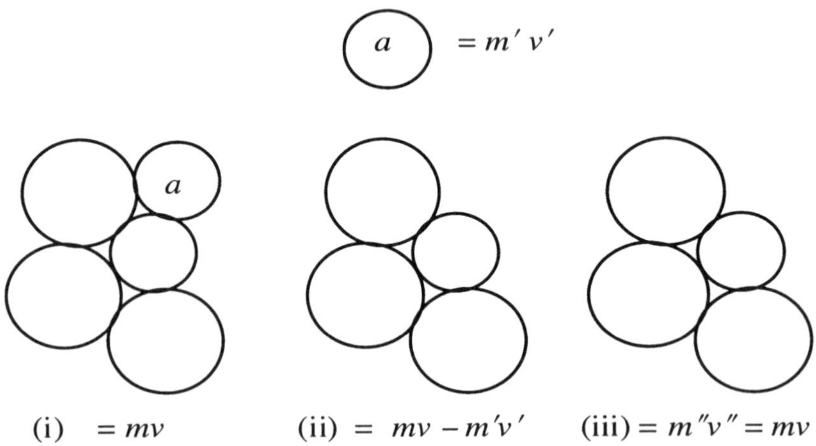

Figure 30. The removal of the body a with quantity of motion $m'v'$ (ii) results in a loss for the system ($= mv - m'v'$) which can only be compensated for by an increase in speed v'' of the remaining bodies m'' (iii); thus, the total quantity of motion prior to the removal (i) will be conserved ($m''v'' - mv$).

Overall, mandating a universal invariant quantity of motion may offer a means of averting Newton's charge that the nature of a plenum precludes the determination of physically important magnitudes (such as velocity, etc.). In order to measure these quantities, of course, a suitable reference frame, or class of such frames, will still be necessary, and cannot be excluded from the specifications of our Cartesian spacetime. Yet, the conservation law guarantees that such reference frames will exist and can be correlated with the reference frame located on the future and past spatial slices. The spacetime can accomplish this task by simply allowing the conservation law itself to "pick out" the appropriate reference frames that preserve the universal magnitude of quantity of motion; a method that we have often labeled as "boot-strapping." That is, given the requirement of an invariant universal quantity, a certain class of material arrangements are automatically chosen, and another excluded, on the future-directed spatial slices. This method assists in linking the information on bodily "tendencies" that exists at an instant with the same information located at the next instant. On each of these spatial slices, consequently, there will always exist a suitable class of reference frames that will measure the desired invariant. The coordinate systems that provide these invariant results will form the group of frames that are relatively non-accelerating (i.e., that are moving at uniform speeds relative to each other).

Another important issue can be resolved by employing the conservation law in the manner suggested, moreover. Our first attempts to construe Descartes' ontology of force—that forces exist at instants, while speed and quantity of motion subsist over time—met with great confusion and uncertainty. Yet, with the assistance of the conservation law, the following picture begins to emerge (as just described): Maintaining an invariant measure of universal quantity of motion, a phenomenon at the level of temporal intervals, regulates and coordinates the behavior of the bodily "tendencies" or forces, a phenomenon at the level of instants. Of course, Descartes would claim that the relationship goes in the other direction; since the forces at an instant cause the motions and conserved quantities that are manifest over time. Either way, this interpretation of Cartesian natural philosophy helps to integrate the morass of seemingly contradictory claims that abound in the *Principles*.

Unfortunately, under-determination problems arise for the Cartesian in this context, as well. Given the material configuration manifest on a single spatial plane, the requirement for a universal invariant magnitude of quantity of motion will on occasion accommodate more than one possible future state. Fixing an invariant quantity of motion, in other words, does not single out a unique future time slice for every arrangement of plenum bodies: thus the conclusion of section 9.3 is similar to the conclusion reached under this interpretation of the KM project. Returning to our gear-analogy, once again, there are numerous configurations of gears, all quite distinct, that will conserve the same quantity of motion (or more likely, in the case of gears, angular momentum). For instance, if given a determinate quantity of motion and a particular arrangement of gears on a single spatial slice, one can rearrange the gears in various ways on the future slices and still retain the same quantity of motion. Therefore, the members of the class of distinct evolutionary states that satisfy the conservation law will be regarded as equally possible outcomes of the material dispositions present at an instant. The additional constraint of an invariant universal magnitude of quantity of motion is thus not sufficient to restore determinism to our Cartesian dynamic theory. This under-determination problem entails, once more, that the information on the velocity or "tendencies" of a single body across a succession of spatial slices will not necessarily supply similar data as regards the remaining inhabitants of the plenum, since many possible future states of the plenum are compatible with that particle's motion.

9.5. Conclusions.

In this chapter, we have attempted to formulate a modern version of Descartes' relationalist physical theory that, in the process of providing the foundation for his conservation law, does not run afoul of Newton's *De gravitatione* argument, problem (2), while also resolving the relational inconsistency in Descartes' treatment of his collision rules, problem (1). In order to achieve this goal, we have examined the kinematics of mechanisms theory for a means of obtaining fixed, or temporarily fixed, reference frames necessary to ground relational measurements of a body's speed, and hence quantity of motion, according to (R2) or (R1*). In this setting, both the implications of Newton's argument and the problems of Cartesian relationalist dynamics/kinematics are exhibited in a new light. As frequently noted, the similarity in detail between KM and Cartesian dynamics is quite striking, especially given Descartes' hypothesis on the configuration of vortices. Both theories advance a view of material interactions that is strongly kinematic, a system or world comprised of rigid bodies locked in harmonious circular movements. On the whole, and despite the obstacles encountered (most notably in section 9.3), utilizing a KM-type method of positing reference frames, in order to solve our problem (1), must rank as one the more successful, and relationally palatable, interpretations or reconstructions of Cartesian natural philosophy—that is, depending on whether or not (R2) or (R1*) are consistent relationalist strategies. It cannot resolve, of course, all the problems raised by Newton's argument, i.e., problem (2): in particular, it was necessary to abandon the search for permanently fixed Φ-landmarks and concentrate, instead, on establishing reference frames at the Φ-landmarks that only endure for finite temporal intervals, or even infinitesimally (although fashioning these latter landmarks proved more problematic; see section 9.3.2).

Finally, returning to a point raised at the beginning of section 9.3.2, it was revealed that one of the implicit assumptions in Newton's argument is a stipulation for some process of conveying information on the states of bodies across time (e.g., position, displacement, velocity, or even "strivings"—the "background geometrical structure" of Part I). If conceived in this fashion, the exploitation of temporary Φ-based reference frames clearly satisfies Newton's conception of the minimum structure required of a theory of space and time. In this version of Cartesian physics, the temporary frames serve as the method of linking across time the information concerning bodily states that exists on the individual time slice. Thus, in a sense, the Φ-based frames are assuming the role normally occupied by absolute/substantival space in the Newtonian picture of the world, as was noted in our previous discussion of the properties of (R2) or (R1*) relationalism. Like absolute space, the temporary reference frames in this relationalist theory allow information on the states of bodies to be meaningfully compared across time.

ENDNOTES

[1] See, e.g., Earman 1989, chapter 2; and, Friedman 1983.

[2] Descartes develops his vortex theory in Pr II and, especially, Pr III of the *Principles*. For a survey of this theory, see E. J. Aiton, *The Vortex Theory of Planetary Motions* (MacDonald: London, 1972).

[3] The relationalist Cartesian spacetime which we will develop essentially forms a member of the broader class of Leibnizian spacetimes, as described in chapter 2.

[4] Descartes clearly viewed space as three-dimensional (AT II 11), but it is unclear if he regarded space as Euclidean. Given the lack of any known alternatives at that time, it is almost certain that he did; yet, Francesco Patrizi and Newton seem to be the only natural philosophers of the period who explicitly described space as Euclidean. See, Grant 1981, 232-234.

[5] The concept of an "enduring geometric feature" is somewhat of a misnomer in our Cartesian spacetime, of course, as will be argued at greater length in the remainder of this chapter. In fact, as mentioned above, there are many ways of mapping Ψ. One could, for instance, select a particle located inside a gear (body) for the mapping and reference frame, although long-term plenum instability will eventually bring the particle to the gear contact surface. Or, a reference frame may be linked to a contact point on an initial time slice, but the frame can

follow one of the points on the gears on each successive slice rather then remain with the contact point. See, Dyson 1969, 38-39.

[6] Although only the non-vectorial quantity speed figures in the conservation law for the quantity of motion, Descartes' natural laws stipulate *directed* uniform motions. Thus a modern Cartesian may attempt to develop a Cartesian concept of velocity, as long as only the scalar property speed figures in the conservation law.

[7] In fact, Descartes' early treatment of the principle of "virtual work" (although not fully equivalent to the modern concept) specifically calls for an infinitesimal measurement of "size (weight) × distance". The virtual work principle is closely related to the quantity of motion, size × speed, and it could be argued that the former is a special instance of the latter—see chapter 5. See, AT II 233-234, and especially, AT II 352-355.

[8] Furthermore, Descartes' definition of "external place", which he defines as the surface of the containing bodies, employs his "surface" concept, which he further reasons must be an abstract notion since "is not a part of one body more than of the other" (Pr II 15). Therefore, it is possible that the contact point between two vortices is an equally abstract concept (since it is the point, or surface, where two such bodies meet).

[9] Presumably, the conserved quantity would be somewhat like our modern concept of angular speed. Also, we are ignoring the effects of friction and material wear, factors that would tend to dissipate the total energy of the system.

CONCLUSION

In providing a final verdict on Cartesian spacetime, we should return to the implications of Newton's argument, the motivating cause behind our lengthy search for a consistent Cartesian theory of space, matter, and motion (and one of the common threads linking the various chapters of the book).

As mentioned at the end of the chapter 1, one can interpret Newton's argument as an appeal to a "background geometrical structure," which we have defined as a means of transferring information across time on bodily states. On a relationalist construal, the relative positions, displacements, and, for Descartes, "tendencies towards motion" of material bodies are what constitutes this information. In short, all of the Cartesian theories we have examined fall within this category; thus Newton's contention, at least on this point, has been vindicated. Many of the Cartesian theories explored in this investigation have simply tied a reference frame to some material portion of the universe to accomplish the task of meaningfully relating information on different spatial slices: in Chapter 8, the reference frames were located at the center-of-mass position between the colliding bodies (a material location, given the plenum); in Chapter 9, the reference frames were located at the contact points between material vortices. In both cases, a material location was chosen as the basis of the reference frames, and hence as the means of conveying information across time, either to overcome the deficiencies in Descartes' collision rules or because such locations were relatively stable.

In chapter 6, section 6.4, and chapter 9, section 9.4, however, an alternative means of transferring information on bodily states was considered. If one were to exploit Descartes' requirement that the quantity of motion be conserved at each instant, then a method of linking the frames thatconserve this quantity on the separate spatial slices can be obtained. Since an invariant quantity of motion is mandated of all spatial slices, there will always exist frames that conserve this quantity, and these frames can be tied to similar coordinate systems on the succeeding instants. Once again, predicting the future dispositions of bodies—or, analogously, uniquely linking the suitable reference frames across spatial slices—is not completely determined on this scheme. Many future configurations of bodies can preserve the same overall quantity of motion. Yet, this method greatly assists in conveying information on the status of material bodies across instantaneous time slices, thus fulfilling Newton's requirement for the construction of a coherent theory of space and time. Despite the ingenuity of this relationalist tactic, it does not violate the "background geometrical structure" supposition, since the conservation law merely supplants the role of absolute space as the means on transferring information on bodily states across time. The success of this project, of course, relies on the viability of the "weaker" forms of relationalism that, as argued strenuously throughout our investigation, Descartes' actual treatment of motion seems to endorse (e.g., the correspondance with Henry More, the non-relational handling of the collision rules, etc.). If the (R2) or (R1*) brands of relationalism are admitted, therefore, the philosophical and conceptual import of Cartesian physics, so long offhandedly dismissed, will rise significantly in stature and importance.

BIBLIOGRAPHY

Aiton, E. J. (1972), *The Vortex Theory of Planetary Motions.* London: MacDonald.

Anderson, W. E. (1976), "Cartesian Motion", in *Motion and Time, Space and Matter: Interrelations in the History and Philosophy of Science,* ed. by P. K. Machamer and R. G. Turnbull. Columbus: Ohio State University Press.

Ariew, R. and A. Gabbey (1998), "The Scholastic background", in *The Cambridge History of Seventeenth Century Philosophy, Vol. 1,* eds., D. Garber and M. Ayers. Cambridge: Cambridge University Press.

Barbour, J. B. (1989), *Absolute or Relative Motion?,* vol. 1, *The Discovery of Dynamics.* Cambridge: Cambridge University Press.

Barbour, J. B., Bertotti, B. (1977),"Gravity and Inertia in a Machian Framework", *Nuovo Cimento* 38B: 1-27.

Beeckman, I. (1939), *Journal,* I, C. de Waard edition. The Hague: M. Nijhoff.

Bennett, J. (1996), "Spinoza's Metaphysics", in *The Cambridge Companion to Spinoza,* ed., D. Garrett. Cambridge: Cambridge U. Press.

Blackwell, R. J. (1963), "Descartes' Concept of Matter", in *The Concept of Matter in Modern Philosophy,* ed., E. McMullin. Notre Dame: U. of Notre Dame Press.

________. (1966), "Descartes' Laws of Motion." *Isis* 57, 220-234.

Brach, R. M. (1991), *Mechanical Impact Dynamics: Rigid Body Collisions.* New York: John Wiley & Sons.

Broughton, J. (1987), "Necessity and Physical Laws in Descartes' Philosophy", *Pacific Philosophical Quarterly,* 68, 205-221.

Brown, G. (1989), "Mathematics, Physics, and Corporeal Substance in Descartes", *Pacific Philosophical Quarterly,* 70 (1989), 281-302.

Brown, J. E. (1978), "The Science of Weights", in *Science in the Middle Ages,* ed., D. C. Lindberg. Chicago: University of Chicago Press.

Clagett, M. (1959), *The Science of Mechanics in Middle Ages.* Madison: University of Wisconsin Press.

Clarke, D. M. (1982), *Descartes' Philosophy of Science.* Manchester: Manchester University Press.

Collins, J. (1971), *Descartes' Philosophy of Nature.* Oxford: Blackwell.

Cordemoy, G. de (1968), *Oeuvres philosophiques.* eds., P. Clair and F. Girbal. Paris: Presses Universitaires de France.

Descartes, R. (1976), *Oeuvres de Descartes,* ed., C. Adams and P. Tannery. Paris: J. Vrin.

________. (1979), *The World,* trans., M. S. Mahoney. New York: Abaris Books.

________. (1983), *Principles of Philosophy,* trans., V. R. Miller and R. P. Miller. Dordrecht: Kluwer Academic Publishers.

________. (1984), *The Philosophical Writings of Descartes, Vol.1,* eds. and trans., J. Cottingham, R. Stoothoff, D. Murdoch. Cambridge: Cambridge University Press.

________. (1984), *The Philosophical Writings of Descartes, Vol.2,* eds. and trans., J. Cottingham, R. Stoothoff, D. Murdoch. Cambridge: Cambridge University Press.

________. (1991), *The Philosophical Writings of Descartes, Vol.3, The Correspondence,* eds. and trans., J. Cottingham, et al. Cambridge: Cambridge University Press.

Damerow, P., et al. (1992), *Exploring the Limits of Preclassical Mechanics.* New York: Springer-Verlag.

Della Rocca, M. (1999), "'If a Body Meets a Body': Descartes on Body-Body Causation", in *New Essays on the Rationalists,* eds., R. J. Gennaro and C. Huenemann. Oxford: Oxford University Press.

Des Chene, D. (1996), *Physiologia: Natural Philosophy in Late Aristotelian and Cartesian Thought*. Ithaca: Cornell University Press.

Dugas, R. (1958), *Mechanics in the Seventeenth Century,* trans., F. Jaquot. Neuchatel: Griffon.

________. (1988), *A History of Mechanics,* trans., J. R. Maddux. New York: Dover Publications.

Dutton, B. (1996), "Indifference, Necessity, and Descartes' Derivation of the Laws of Motion", *Journal of the History of Philosophy,* 34, 193-212.

Dyson, A. (1969), *A General Theory of the Kinematics and Geometry of Gears in Three Dimensions*. Oxford: Clarendon Press.

Earman, J. (1989), *World Enough and Space-Time*. Cambridge, MA: MIT Press.

Elzinga, A. (1972), *On A Research Program in Early Modern Physics*. New York: Humanities Press.

Friedman, M. (1983) *Foundations of Space-Time Theories*. Princeton: Princeton University Press.

Gabbey, A. (1973), "Essay review of W. L. Scott's, *The Conflict Between Atomism and Conservation Theory: 1644-1860", Studies in History and Philosophy of Science,* 3, 373-385.

________. (1980), "Force and Inertia in the Seventeenth Century: Descartes and Newton." in *Descartes: Philosophy, Mathematics and Physics,* ed., S. Gaukroger. Sussex: Harvester Press.

________. (1980a), "Huygens and Mechanics", in *Studies on Christiaan Huygens,* H. J. M. Bos et al, eds., Swetz & Zeitlinger, Linse, 166-199.

________. (1985), "The Mechanical Philosophy and its Problems: Mechanical Explanations, Impenetrability, and Perpetual Motion", in *Change and Progress in Modern Science,* ed., J. C. Pitt. Dordrecht: D. Reidel.

________. (1993), "Descartes's Physics and Descartes's Mechanics: Chicken and Egg", in *Essays on the Philosophy and Science of René Descartes,* ed., S. Voss. Oxford: Oxford University Press.

Galilei, G. (1957), *Letters on the Sunspots,* in *Discoveries and Opinions of Galileo,* trans., S. Drake. New York: Anchor.

________. (1960), *Discourse on Bodies in Water,* trans., T. Salusbury, ed., S. Drake. Urbana: University of Illinois Press.

Garber, D. (1992), *Descartes' Metaphysical Physics.* Chicago: University of Chicago Press.

Gaukroger, S. (1995), *Descartes: An Intellectual Biography.* Oxford: Oxford University Press.

Gaukroger, S. (2000), "The Foundational Role of Hydrostatics in Descartes' Natural Philosophy" in *Descartes' Natural Philosophy,* eds., S. Gaukroger, S. Schuster, S. Sutton. London: Routledge.

Gueroult, M. (1980), "The Metaphysics and Physics of Force in Descartes", in *Descartes: Philosophy, Mathematics and Physics,* ed., S. Gaukroger. Sussex: Harvester Press.

________. (1984), *Descartes' Philosophy Interpreted According to the Order of Reasons, Vol. II,* trans., R. Ariew. Minneapolis: U. of Minnesota Press

Grant, E. (1981), *Much Ado About Nothing: Theories of Space and Vacuum from the Middle Ages to the Scientific Revolution.* Cambridge: Cambridge University Press.

________. (1994), *Planets, Stars, & Orbs: The Medieval Cosmos, 1200-1687.* Cambridge: Cambridge University Press.

Grene, M. (1985), *Descartes.* Minneapolis: U. of Minnesota Press.

Grosholz, E. (1991), *Cartesian Method and the Problem of Reduction.* Oxford: Clarendon Press.

________. (1994), "Descartes and the Individuation of Physical Objects", in *Individuation and Identity in Early Modern Philosophy,* ed., K. F. Barber and J. J. E. Gracia. Albany: State U. of New York Press.

Hall, H. A. (1976), "Huygens and Newton", in *The Anglo-Dutch Contribution to the Civilization of Early Modern Society,* eds., C. Wilson et al. London: Oxford University Press.

Hatfield, G. (1979), "Force (God) in Descartes' Physics", in *Studies in History and Philosophy of Science,* 10: 113-140.

Herival, J. (1965), *The Background to Newton's 'Principia'*. Oxford: Oxford University Press.

Hoffman, P. (1999), "The Unity of Descartes' Man", in *The Rationalists,* ed., D. Pereboom. Lanham: Rowman & Littlefield.

Hooper, W. (1998), "Inertial Problems in Galileo's Preinertial Framework", in *The Cambridge Companion to Galileo,* ed., P. Machamer. Cambridge: Cambridge University Press.

Hübner, K. (1983), *Critique of Scientific Reason,* trans. by P. R. Dixon and H. M. Dixon. Chicago: University of Chicago Press.

Huygens, C. (1950), *Oeuvres Complètes.* La Haye: Société Hollandaise des Sciences.

Jammer, M. (1961), *Concepts of Mass in Classical and Modern Physics.* Cambridge, Mass.: Harvard University Press.

________. (1969), *Concepts of Space,* 2nd ed. Cambridge, MA: Harvard U. Press.

Keeling, S. V. (1968), *Descartes,* 2nd ed. Oxford: Oxford U. Press.

Koyré, A. (1978), *Galileo Studies.* Trans. by J. Mepham. Atlantic Highlands, N.J.:Humanities Press.

Lawden, D. F. (1982), *An Introduction to Tensor Calculus, Relativity and Cosmology.* 3rd ed. Chichester: John Wiley & Sons.

Laymon, R. (1978), "Newton's Bucket Experiment", *Journal of the History of Philosophy,* 16, 399-413.

Leibniz, G. W. (1970), *G. W. Leibniz: Philosophical Papers and Letters,* trans. and ed., L. E. Loemker. Dordrecht: D. Reidel.

Lennon, T. M. (1993), *The Battle of the Gods and Giants: The Legacy of Descartes and Gassendi*. Princeton: Princeton University Press.

Love, A. E. H. (1944), *A Treatise on the Mathematical Theory of Elasticity*. New York: Dover.

Lucas, J. R. (1984), *Space, Time, and Causality*. Oxford: Oxford University Press.

MacGuire, J. E. (1983), "Space, Geometrical Objects and Infinity: Newton and Descartes on Extension", in *Nature Mathematized: Historical and Philosophical Case Studies in Classical Modern Natural Philosophy,* ed., W. R. Shea. Dordrecht: D. Reidel.

Mach, E. (1942), *The Science of Mechanics,* 9th ed. London: Open Court.

Machamer, P. K. (1976), "Causality and Explanation in Descartes' Natural Philosophy", in *Motion and Time, Space and Matter: Interrelations in the History and Philosophy of Science,* eds., P. K. Machamer and R. G. Turnbull. Columbus: Ohio State University Press.

Maier, A. (1982), *On the Threshold of Exact Science,* trans., S. D. Sargent. Philadelphia: U. of Pennsylvania Press.

Mancosu, P. (1996), *Philosophy of Mathematics and Mathematical Practice in the Seventeenth Century*. Oxford: Oxford University Press.

Marshall, jr., D. (1979), *Prinzipien der Descartes: Exegese*. Freiburg: Verlag Karl Alber.

Mase, G. E. (1970), *Theory and Problems of Continuum Mechanics*. New York: McGraw-Hill.

McLaughlin, P. (2000), "Force, Determination, Impact", in *Descartes' Natural Philosophy,* eds., S. Gaukroger, S. Schuster, S. Sutton. London: Routledge.

Merchant, C. (1980), *The Death of Nature: Women, Ecology, and the Scientific Revolution*. New York: HarperCollins Pub.

Misner, C. W., Thorne, K. S., Wheeler, J. A. (1973), *Gravitation*. San Francisco: W. H. Freeman.

Murdoch, J. E., and E. D. Sylla (1978), "The Science of Motion", in *Science in the Middle Ages,* ed., D. C. Lindberg. Chicago: University of Chicago Press.

Nadler, S. (1990), "Deduction, Confirmation, and the Laws of Nature in Descartes' *Principia Philosophiae." Journal of the History of Philosophy,* 28, 359-383.

Neile, W. (1968), "Hypothesis of Motion," in *The Correspondence of Henry Oldenburg, Vol. 5,* eds., A. R. Hall and M. B. Hall. Madison: University of Wisconsin Press.

Nelson, A. (1995), "Micro-Chaos and Idealization in Cartesian Physics", *Philosophical Studies,* 77, 377-391.

Newton, I. (1962a), *De gravitatione et aequipondio fluidorum,* in *Unpublished Scientific Papers of Isaac Newton,* trans. and eds., A. R. Hall and M. B. Hall. Cambridge: Cambridge University Press.

________. (1962b), *Mathematical Principals of Natural Philosophy,* trans., A. Motte and F. Cajori. Berkeley: University of California Press.

Normore, C. (1993), "The Necessity in Deduction: Cartesian Inference and its Medieval Background." *Synthese,* 96, 437-454.

Norton, J. D. (1992), "Philosophy of Space and Time", in *Introduction to the Philosophy of Science,* eds., M. H. Salmon, et al. Englewood Cliffs: Prentice Hall.

Prendergast, T. (1972), "Descartes and the Relativity of Motion." *The Modern Schoolman,* 49, 64-72.

________. (1975), "Motion, Action, and Tendency in Descartes' Physics", *Journal of the History of Philosophy,* 13, 453-462.

(Rodis-)Lewis, G. (1950), *L'individualité selon Descartes.* Paris: J. Vrin.

Rodis-Lewis, G. (1971) *L'oeuvre de Descartes.* Paris: J. Vrin.

Rohault, J. (1969), *A System of Natural Philosophy, vol. 1,* trans., J. Clarke and S. Clarke, 1723. New York: Johnson Reprint Corp.

Schuster, J. A. (1980), "Descartes' *Mathesis Universalis: 1619-28"*, in *Descartes: Philosophy, Mathematics and Physics,* ed., S. Gaukroger. Sussex: Harvester Press.

Scott, W. L. (1970), *The Conflict Between Atomism and Conservation Theory 1644-1860.* London: MacDonald.

Shapiro, A. E. (1980), "Huygens' Kinematic Theory of Light", in *Studies on Christiaan Huygens,* eds., H. J. M. Bos et al. Linse: Swetz & Zeitlinger.

Shea, W. R. (1991), *The Magic of Numbers and Motion: The Scientific Career of René Descartes.* Canton, Mass.: Science History Publications.

Sklar, L. (1974), *Space, Time, and Spacetime.* U. of California Press, Berkeley.

Slowik, E. (1996), "Perfect Solidity: Natural Laws and The Problem of Matter in Descartes' Universe." *History of Philosophy Quarterly,* 13.

_______. (1997), "Huygens' Center-of-Mass Space-Time Reference Frame: Constructing a Cartesian Dynamics in The Wake of Newton's *'De gravitatione'* Argument", *Synthese* 112: 247-269.

_______. (1998) "Descartes, Space-Time, and Relational Motion." *Philosophy of Science* 66: 117-140.

Snelders, H. A. M. (1980), "Christiaan Huygens and the concepts of matter", in *Studies on Christiaan Huygens,* H. J. M. Bos et al, eds., Swetz & Zeitlinger, Linse, 104-125.

Sorabji, R. (1988), *Matter, Space, and Motion.* Cornell University Press, Ithaca.

Spinoza, B. (1925), *Spinoza Opera,* ed., G. Gerhardt. Heidelberg: C. Winter.

Stein, H. (1967), "Newtonian Space-Time", *Texas Quarterly,* 10, 174-200.

________. (1977), "Some Pre-History of General Relativity", in *Foundations of Space-Time Theories,* J. Earman, C. Glymour, and J. Stachel, eds., Minnesota Studies in the Philosophy of Science, vol. 8, University of Minnesota Press, Minneapolis, 3-49.

Truesdell, C. (1984), *An Idiot's Fugitive Essays on Science*. New York: Springer Verlag.

Westfall, R. (1971), *The Concept of Force in Newton's Physics*. London: MacDonald.

Westman, R. S. (1980), "Huygens and the Problem of Cartesianism", in *Studies of Christiaan Huygens,* eds., H. J. M. Bos et al. Linse: Swetz & Zeitlinger.

Wilson, M. (1989), "Critical Notice: John Earman's *A Primer on Determinism*", *Philosophy of Science,* 56.

________. (1993), "There's a Hole and a Bucket, Dear Leibniz", in *Midwest Studies in Philosophy Vol. XVIII, Philosophy of Science,* eds., P. A. French, T. E. Uehling, Jr., H. K. Wettstein. Notre Dame, IN: U. of Notre Dame Press.

Woolhouse, R. S. (1993), *Descartes, Spinoza, Leibniz: The Concept of Substance in Seventeenth Century Metaphysics*. London: Routledge.

Zimmerman, J. R. (1962), *Elementary Kinematics of Mechanisms*. New York: John Wiley & Sons.

INDEX

ARCHIVES INTERNATIONALES D'HISTOIRE DES IDÉES

*

INTERNATIONAL ARCHIVES OF THE HISTORY OF IDEAS

1. E. Labrousse: *Pierre Bayle.* Tome I: *Du pays de foix à la cité d'Erasme.* 1963; 2nd printing 1984 ISBN 90-247-3136-4
   For Tome II *see below under Volume 6.*
2. P. Merlan: *Monopsychism, Mysticism, Metaconsciousness.* Problems of the Soul in the Neoaristotelian and Neoplatonic Tradition. 1963; 2nd printing 1969 ISBN 90-247-0178-3
3. H.G. van Leeuwen: *The Problem of Certainty in English Thought, 1630–1690.* With a Preface by R.H. Popkin. 1963; 2nd printing 1970 ISBN 90-247-0179-1
4. P.W. Janssen: *Les origines de la réforme des Carmes en France au 17e Siècle.* 1963; 2nd printing 1969 ISBN 90-247-0180-5
5. G. Sebba: *Bibliographia Cartesiana.* A Critical Guide to the Descartes Literature (1800–1960). 1964 ISBN 90-247-0181-3
6. E. Labrousse: *Pierre Bayle.* Tome II: *Heterodoxie et rigorisme.* 1964 ISBN 90-247-0182-1
7. K.W. Swart: *The Sense of Decadence in 19th-Century France.* 1964 ISBN 90-247-0183-X
8. W. Rex: *Essays on Pierre Bayle and Religious Controversy.* 1965 ISBN 90-247-0184-8
9. E. Heier: *L.H. Nicolay (1737–1820) and His Contemporaries.* Diderot, Rousseau, Voltaire, Gluck, Metastasio, Galiani, D'Escherny, Gessner, Bodmer, Lavater, Wieland, Frederick II, Falconet, W. Robertson, Paul I, Cagliostro, Gellert, Winckelmann, Poinsinet, Lloyd, Sanchez, Masson, and Others. 1965 ISBN 90-247-0185-6
10. H.M. Bracken: *The Early Reception of Berkeley's Immaterialism, 1710–1733.* [1958] Rev. ed. 1965 ISBN 90-247-0186-4
11. R.A. Watson: *The Downfall of Cartesianism, 1673–1712.* A Study of Epistemological Issues in Late 17th-Century Cartesianism. 1966 ISBN 90-247-0187-2
12. R. Descartes: *Regulæ ad Directionem Ingenii.* Texte critique établi par Giovanni Crapulli avec la version hollandaise du 17e siècle. 1966 ISBN 90-247-0188-0
13. J. Chapelain: *Soixante-dix-sept Lettres inédites à Nicolas Heinsius (1649–1658).* Publiées d'après le manuscrit de Leyde avec une introduction et des notes par B. Bray. 1966 ISBN 90-247-0189-9
14. C. B. Brush: *Montaigne and Bayle.* Variations on the Theme of Skepticism. 1966 ISBN 90-247-0190-2
15. B. Neveu: *Un historien à l'Ecole de Port-Royal.* Sébastien le Nain de Tillemont (1637–1698). 1966 ISBN 90-247-0191-0
16. A. Faivre: *Kirchberger et l'Illuminisme du 18e siècle.* 1966 ISBN 90-247-0192-9
17. J.A. Clarke: *Huguenot Warrior.* The Life and Times of Henri de Rohan (1579–1638). 1966 ISBN 90-247-0193-7
18. S. Kinser: *The Works of Jacques-Auguste de Thou.* 1966 ISBN 90-247-0194-5
19. E.F. Hirsch: *Damião de Gois.* The Life and Thought of a Portuguese Humanist (1502–1574). 1967 ISBN 90-247-0195-3
20. P.J.S. Whitemore: *The Order of Minims in 17th-Century France.* 1967 ISBN 90-247-0196-1
21. H. Hillenaar: *Fénelon et les Jésuites.* 1967 ISBN 90-247-0197-X
22. W.N. Hargreaves-Mawdsley: *The English Della Cruscans and Their Time, 1783–1828.* 1967 ISBN 90-247-0198-8
23. C.B. Schmitt: *Gianfrancesco Pico della Mirandola (1469–1533) and his Critique of Aristotle.* 1967 ISBN 90-247-0199-6
24. H.B. White: *Peace among the Willows.* The Political Philosophy of Francis Bacon. 1968 ISBN 90-247-0200-3

ARCHIVES INTERNATIONALES D'HISTOIRE DES IDÉES
*
INTERNATIONAL ARCHIVES OF THE HISTORY OF IDEAS

25. L. Apt: *Louis-Philippe de Ségur.* An Intellectual in a Revolutionary Age. 1969 ISBN 90-247-0201-1
26. E.H. Kadler: *Literary Figures in French Drama (1784–1834).* 1969 ISBN 90-247-0202-X
27. G. Postel: *Le Thrésor des prophéties de l'univers.* Manuscrit publié avec une introduction et des notes par F. Secret. 1969 ISBN 90-247-0203-8
28. E.G. Boscherini: *Lexicon Spinozanum.* 2 vols., 1970 Set ISBN 90-247-0205-4
29. C.A. Bolton: *Church Reform in 18th-Century Italy.* The Synod of Pistoia (1786). 1969 ISBN 90-247-0208-9
30. D. Janicaud: *Une généalogie du spiritualisme français.* Aux sources du bergsonisme: [Félix] Ravaisson [1813–1900] et la métaphysique. 1969 ISBN 90-247-0209-7
31. J.-E. d'Angers: *L'Humanisme chrétien au 17e siècle.* St. François de Sales et Yves de Paris. 1970 ISBN 90-247-0210-0
32. H.B. White: *Copp'd Hills towards Heaven.* Shakespeare and the Classical Polity. 1970 ISBN 90-247-0250-X
33. P.J. Olscamp: *The Moral Philosophy of George Berkeley.* 1970 ISBN 90-247-0303-4
34. C.G. Noreña: *Juan Luis Vives (1492–1540).* 1970 ISBN 90-247-5008-3
35. J. O'Higgens: *Anthony Collins (1676–1729), the Man and His World.* 1970 ISBN 90-247-5007-5
36. F.T. Brechka: *Gerard van Swieten and His World (1700–1772).* 1970 ISBN 90-247-5009-1
37. M.H. Waddicor: *Montesquieu and the Pilosophy of Natural Law.* 1970 ISBN 90-247-5039-3
38. O.R. Bloch: *La Philosophie de Gassendi (1592–1655).* Nominalisme, matérialisme et métaphysique. 1971 ISBN 90-247-5035-0
39. J. Hoyles: *The Waning of the Renaissance (1640–1740).* Studies in the Thought and Poetry of Henry More, John Norris and Isaac Watts. 1971 ISBN 90-247-5077-6
    For Henry More, *see also below under Volume 122 and 127.*
40. H. Bots: *Correspondance de Jacques Dupuy et de Nicolas Heinsius (1646–1656).* 1971 ISBN 90-247-5092-X
41. W.C. Lehmann: *Henry Home, Lord Kames, and the Scottish Enlightenment.* A Study in National Character and in the History of Ideas. 1971 ISBN 90-247-5018-0
42. C. Kramer: *Emmery de Lyere et Marnix de Sainte Aldegonde.* Un admirateur de Sébastien Franck et de Montaigne aux prises avec le champion des calvinistes néerlandais.[Avec le texte d'Emmery de Lyere:] *Antidote ou contrepoison contre les conseils sanguinaires et envinemez de Philippe de Marnix Sr. de Ste. Aldegonde.* 1971 ISBN 90-247-5136-5
43. P. Dibon: *Inventaire de la correspondance (1595–1650) d'André Rivet (1572–1651).* 1971 ISBN 90-247-5112-8
44. K.A. Kottman: *Law and Apocalypse.* The Moral Thought of Luis de Leon (1527?–1591). 1972 ISBN 90-247-1183-5
45. F.G. Nauen: *Revolution, Idealism and Human Freedom.* Schelling, Hölderlin and Hegel, and the Crisis of Early German Idealism. 1971 ISBN 90-247-5117-9
46. H. Jensen: *Motivation and the Moral Sense in Francis Hutcheson's* [1694–1746] *Ethical Theory.* 1971 ISBN 90-247-1187-8
47. A. Rosenberg: *[Simon] Tyssot de Patot and His Work (1655–1738).* 1972 ISBN 90-247-1199-1
48. C. Walton: *De la recherche du bien.* A study of [Nicolas de] Malebranche's [1638–1715] Science of Ethics. 1972 ISBN 90-247-1205-X

ARCHIVES INTERNATIONALES D'HISTOIRE DES IDÉES
*
INTERNATIONAL ARCHIVES OF THE HISTORY OF IDEAS

49. P.J.S. Whitmore (ed.): *A 17th-Century Exposure of Superstition.* Select Text of Claude Pithoys (1587–1676). 1972 ISBN 90-247-1298-X
50. A. Sauvy: *Livres saisis à Paris entre 1678 et 1701.* D'après une étude préliminaire de Motoko Ninomiya. 1972 ISBN 90-247-1347-1
51. W.R. Redmond: *Bibliography of the Philosophy in the Iberian Colonies of America.* 1972 ISBN 90-247-1190-8
52. C.B. Schmitt: *Cicero Scepticus.* A Study of the Influence of the *Academica* in the Renaissance. 1972 ISBN 90-247-1299-8
53. J. Hoyles: *The Edges of Augustanism.* The Aesthetics of Spirituality in Thomas Ken, John Byrom and William Law. 1972 ISBN 90-247-1317-X
54. J. Bruggeman and A.J. van de Ven (éds.): *Inventaire* des pièces d'Archives françaises se rapportant à l'Abbaye de Port-Royal des Champs et son cercle et à la Résistance contre la Bulle *Unigenitus* et à l'Appel. 1972 ISBN 90-247-5122-5
55. J.W. Montgomery: *Cross and Crucible.* Johann Valentin Andreae (1586–1654), Phoenix of the Theologians. Volume I: Andreae's Life, World-View, and Relations with Rosicrucianism and Alchemy; Volume II: The *Chymische Hochzeit* with Notes and Commentary. 1973 Set ISBN 90-247-5054-7
56. O. Lutaud: *Des révolutions d'Angleterre à la Révolution française.* Le tyrannicide & *Killing No Murder* (Cromwell, *Athalie*, Bonaparte). 1973 ISBN 90-247-1509-1
57. F. Duchesneau: *L'Empirisme de Locke.* 1973 ISBN 90-247-1349-8
58. R. Simon (éd.): *Henry de Boulainviller* – Œuvres Philosophiques, Tome I. 1973 ISBN 90-247-1332-3

    For Œuvres Philosophiques, Tome II *see below under Volume 70.*
59. E.E. Harris: *Salvation from Despair.* A Reappraisal of Spinoza's Philosophy. 1973 ISBN 90-247-5158-6
60. J.-F. Battail: *L'Avocat philosophe Géraud de Cordemoy (1626–1684).* 1973 ISBN 90-247-1542-3
61. T. Liu: *Discord in Zion.* The Puritan Divines and the Puritan Revolution (1640–1660). 1973 ISBN 90-247-5156-X
62. A. Strugnell: *Diderot's Politics.* A Study of the Evolution of Diderot's Political Thought after the *Encyclopédie.* 1973 ISBN 90-247-1540-7
63. G. Defaux: *Pantagruel et les Sophistes.* Contribution à l'histoire de l'humanisme chrétien au 16e siècle. 1973 ISBN 90-247-1566-0
64. G. Planty-Bonjour: *Hegel et la pensée philosophique en Russie (1830–1917).* 1974 ISBN 90-247-1576-8
65. R.J. Brook: [*George*] *Berkeley's Philosophy of Science.* 1973 ISBN 90-247-1555-5
66. T.E. Jessop: *A Bibliography of George Berkeley.* With: *Inventory of Berkeley's Manuscript Remains* by A.A. Luce. 2nd revised and enlarged ed. 1973 ISBN 90-247-1577-6
67. E.I. Perry: *From Theology to History.* French Religious Controversy and the Revocation of the Edict of Nantes. 1973 ISBN 90-247-1578-4
68. P. Dibbon, H. Bots et E. Bots-Estourgie: *Inventaire de la correspondance (1631–1671) de Johannes Fredericus Gronovius* [1611–1671]. 1974 ISBN 90-247-1600-4
69. A.B. Collins: *The Secular is Sacred.* Platonism and Thomism in Marsilio Ficino's *Platonic Theology.* 1974 ISBN 90-247-1588-1

ARCHIVES INTERNATIONALES D'HISTOIRE DES IDÉES
*
INTERNATIONAL ARCHIVES OF THE HISTORY OF IDEAS

70. R. Simon (éd.): *Henry de Boulainviller.* Œuvres Philosophiques, Tome II. 1975 ISBN 90-247-1633-0
For Œuvres Philosophiques, Tome I *see under Volume 58.*
71. J.A.G. Tans et H. Schmitz du Moulin: *Pasquier Quesnel devant la Congrégation de l'Index.* Correspondance avec Francesco Barberini et mémoires sur la mise à l'Index de son édition des Œuvres de Saint Léon, publiés avec introduction et annotations. 1974 ISBN 90-247-1661-6
72. J.W. Carven: *Napoleon and the Lazarists (1804–1809).* 1974 ISBN 90-247-1667-5
73. G. Symcox: *The Crisis of French Sea Power (1688–1697).* From the *Guerre d'Escadre* to the *Guerre de Course.* 1974 ISBN 90-247-1645-4
74. R. MacGillivray: *Restoration Historians and the English Civil War.* 1974 ISBN 90-247-1678-0
75. A. Soman (ed.): *The Massacre of St. Bartholomew.* Reappraisals and Documents. 1974 ISBN 90-247-1652-7
76. R.E. Wanner: *Claude Fleury (1640–1723) as an Educational Historiographer and Thinker.* With an Introduction by W.W. Brickman. 1975 ISBN 90-247-1684-5
77. R.T. Carroll: *The Common-Sense Philosophy of Religion of Bishop Edward Stillingfleet (1635–1699).* 1975 ISBN 90-247-1647-0
78. J. Macary: *Masque et lumières au 18e [siècle].* André-François Deslandes, Citoyen et philosophe (1689–1757). 1975 ISBN 90-247-1698-5
79. S.M. Mason: *Montesquieu's Idea of Justice.* 1975 ISBN 90-247-1670-5
80. D.J.H. van Elden: *Esprits fins et esprits géométriques dans les portraits de Saint-Simon.* Contributions à l'étude du vocabulaire et du style. 1975 ISBN 90-247-1726-4
81. I. Primer (ed.): *Mandeville Studies.* New Explorations in the Art and Thought of Dr Bernard Mandeville (1670–1733). 1975 ISBN 90-247-1686-1
82. C.G. Noreña: *Studies in Spanish Renaissance Thought.* 1975 ISBN 90-247-1727-2
83. G. Wilson: *A Medievalist in the 18th Century.* Le Grand d'Aussy and the Fabliaux ou Contes. 1975 ISBN 90-247-1782-5
84. J.-R. Armogathe: *Theologia Cartesiana.* L'explication physique de l'Eucharistie chez Descartes et Dom Robert Desgabets. 1977 ISBN 90-247-1869-4
85. Bérault Stuart, Seigneur d'Aubigny: *Traité sur l'art de la guerre.* Introduction et édition par Élie de Comminges. 1976 ISBN 90-247-1871-6
86. S.L. Kaplan: *Bread, Politics and Political Economy in the Reign of Louis XV.* 2 vols., 1976 Set ISBN 90-247-1873-2
87. M. Lienhard (ed.): *The Origins and Characteristics of Anabaptism / Les débuts et les caractéristiques de l'Anabaptisme.* With an Extensive Bibliography / Avec une bibliographie détaillée. 1977 ISBN 90-247-1896-1
88. R. Descartes: *Règles utiles et claires pour la direction de l'esprit en la recherche de la vérité.* Traduction selon le lexique cartésien, et annotation conceptuelle par J.-L. Marion. Avec des notes mathématiques de P. Costabel. 1977 ISBN 90-247-1907-0
89. K. Hardesty: *The 'Supplément' to the 'Encyclopédie'.* [Diderot et d'Alembert]. 1977 ISBN 90-247-1965-8
90. H.B. White: *Antiquity Forgot.* Essays on Shakespeare, [Francis] Bacon, and Rembrandt. 1978 ISBN 90-247-1971-2
91. P.B.M. Blaas: *Continuity and Anachronism.* Parliamentary and Constitutional Development in Whig Historiography and in the Anti-Whig Reaction between 1890 and 1930. 1978 ISBN 90-247-2063-X

ARCHIVES INTERNATIONALES D'HISTOIRE DES IDÉES

*

INTERNATIONAL ARCHIVES OF THE HISTORY OF IDEAS

---

92. S.L. Kaplan (ed.): *La Bagarre.* Ferdinando Galiani's (1728–1787) 'Lost' Parody. With an Introduction by the Editor. 1979 ISBN 90-247-2125-3
93. E. McNiven Hine: *A Critical Study of [Étienne Bonnot de] Condillac's* [1714–1780] *'Traité des Systèmes'.* 1979 ISBN 90-247-2120-2
94. M.R.G. Spiller: *Concerning Natural Experimental Philosphy.* Meric Casaubon [1599–1671] and the Royal Society. 1980 ISBN 90-247-2414-7
95. F. Duchesneau: *La physiologie des Lumières.* Empirisme, modèles et théories. 1982 ISBN 90-247-2500-3
96. M. Heyd: *Between Orthodoxy and the Enlightenment.* Jean-Robert Chouet [1642–1731] and the Introduction of Cartesian Science in the Academy of Geneva. 1982 ISBN 90-247-2508-9
97. James O'Higgins: *Yves de Vallone* [1666/7–1705]: *The Making of an Esprit Fort.* 1982 ISBN 90-247-2520-8
98. M.L. Kuntz: *Guillaume Postel* [1510–1581]. Prophet of the Restitution of All Things. His Life and Thought. 1981 ISBN 90-247-2523-2
99. A. Rosenberg: *Nicolas Gueudeville and His Work (1652–172?).* 1982 ISBN 90-247-2533-X
100. S.L. Jaki: *Uneasy Genius: The Life and Work of Pierre Duhem* [1861-1916]. 1984 ISBN 90-247-2897-5; Pb (1987) 90-247-3532-7
101. Anne Conway [1631–1679]: *The Principles of the Most Ancient Modern Philosophy.* Edited and with an Introduction by P. Loptson. 1982 ISBN 90-247-2671-9
102. E.C. Patterson: *[Mrs.] Mary [Fairfax Greig] Sommerville* [1780–1872] *and the Cultivation of Science (1815–1840).* 1983 ISBN 90-247-2823-1
103. C.J. Berry: *Hume, Hegel and Human Nature.* 1982 ISBN 90-247-2682-4
104. C.J. Betts: *Early Deism in France.* From the so-called 'déistes' of Lyon (1564) to Voltaire's 'Lettres philosophiques' (1734). 1984 ISBN 90-247-2923-8
105. R. Gascoigne: *Religion, Rationality and Community.* Sacred and Secular in the Thought of Hegel and His Critics. 1985 ISBN 90-247-2992-0
106. S. Tweyman: *Scepticism and Belief in Hume's 'Dialogues Concerning Natural Religion'.* 1986 ISBN 90-247-3090-2
107. G. Cerny: *Theology, Politics and Letters at the Crossroads of European Civilization.* Jacques Basnage [1653–1723] and the Baylean Huguenot Refugees in the Dutch Republic. 1987 ISBN 90-247-3150-X
108. Spinoza's *Algebraic Calculation of the Rainbow & Calculation of Changes.* Edited and Translated from Dutch, with an Introduction, Explanatory Notes and an Appendix by M.J. Petry. 1985 ISBN 90-247-3149-6
109. R.G. McRae: *Philosophy and the Absolute.* The Modes of Hegel's Speculation. 1985 ISBN 90-247-3151-8
110. J.D. North and J.J. Roche (eds.): *The Light of Nature.* Essays in the History and Philosophy of Science presented to A.C. Crombie. 1985 ISBN 90-247-3165-8
111. C. Walton and P.J. Johnson (eds.): *[Thomas] Hobbes's 'Science of Natural Justice'.* 1987 ISBN 90-247-3226-3
112. B.W. Head: *Ideology and Social Science.* Destutt de Tracy and French Liberalism. 1985 ISBN 90-247-3228-X
113. A.Th. Peperzak: *Philosophy and Politics.* A Commentary on the Preface to Hegel's *Philosophy of Right.* 1987 ISBN Hb 90-247-3337-5; Pb ISBN 90-247-3338-3

ARCHIVES INTERNATIONALES D'HISTOIRE DES IDÉES
*
INTERNATIONAL ARCHIVES OF THE HISTORY OF IDEAS

114. S. Pines and Y. Yovel (eds.): *Maimonides* [1135-1204] *and Philosophy.* Papers Presented at the 6th Jerusalem Philosophical Encounter (May 1985). 1986 ISBN 90-247-3439-8
115. T.J. Saxby: *The Quest for the New Jerusalem, Jean de Labadie* [1610–1674] *and the Labadists (1610–1744).* 1987 ISBN 90-247-3485-1
116. C.E. Harline: *Pamphlets, Printing, and Political Culture in the Early Dutch Republic.* 1987 ISBN 90-247-3511-4
117. R.A. Watson and J.E. Force (eds.): *The Sceptical Mode in Modern Philosophy.* Essays in Honor of Richard H. Popkin. 1988 ISBN 90-247-3584-X
118. R.T. Bienvenu and M. Feingold (eds.): *In the Presence of the Past.* Essays in Honor of Frank Manuel. 1991 ISBN 0-7923-1008-X
119. J. van den Berg and E.G.E. van der Wall (eds.): *Jewish-Christian Relations in the 17th Century.* Studies and Documents. 1988 ISBN 90-247-3617-X
120. N. Waszek: *The Scottish Enlightenment and Hegel's Account of 'Civil Society'.* 1988 ISBN 90-247-3596-3
121. J. Walker (ed.): *Thought and Faith in the Philosophy of Hegel.* 1991 ISBN 0-7923-1234-1
122. Henry More [1614–1687]: *The Immortality of the Soul.* Edited with Introduction and Notes by A. Jacob. 1987 ISBN 90-247-3512-2
123. P.B. Scheurer and G. Debrock (eds.): *Newton's Scientific and Philosophical Legacy.* 1988 ISBN 90-247-3723-0
124. D.R. Kelley and R.H. Popkin (eds.): *The Shapes of Knowledge from the Renaissance to the Enlightenment.* 1991 ISBN 0-7923-1259-7
125. R.M. Golden (ed.): *The Huguenot Connection.* The Edict of Nantes, Its Revocation, and Early French Migration tô South Carolina. 1988 ISBN 90-247-3645-5
126. S. Lindroth: *Les chemins du savoir en Suède.* De la fondation de l'Université d'Upsal à Jacob Berzelius. Études et Portraits. Traduit du suédois, présenté et annoté par J.-F. Battail. Avec une introduction sur Sten Lindroth par G. Eriksson. 1988 ISBN 90-247-3579-3
127. S. Hutton (ed.): *Henry More (1614–1687). Tercentenary Studies.* With a Biography and Bibliography by R. Crocker. 1989 ISBN 0-7923-0095-5
128. Y. Yovel (ed.): *Kant's Practical Philosophy Reconsidered.* Papers Presented at the 7th Jerusalem Philosophical Encounter (December 1986). 1989 ISBN 0-7923-0405-5
129. J.E. Force and R.H. Popkin: *Essays on the Context, Nature, and Influence of Isaac Newton's Theology.* 1990 ISBN 0-7923-0583-3
130. N. Capaldi and D.W. Livingston (eds.): *Liberty in Hume's 'History of England'.* 1990 ISBN 0-7923-0650-3
131. W. Brand: *Hume's Theory of Moral Judgment.* A Study in the Unity of *A Treatise of Human Nature.* 1992 ISBN 0-7923-1415-8
132. C.E. Harline (ed.): *The Rhyme and Reason of Politics in Early Modern Europe.* Collected Essays of Herbert H. Rowen. 1992 ISBN 0-7923-1527-8
133. N. Malebranche: *Treatise on Ethics* (1684). Translated and edited by C. Walton. 1993 ISBN 0-7923-1763-7
134. B.C. Southgate: *'Covetous of Truth'.* The Life and Work of Thomas White (1593–1676). 1993 ISBN 0-7923-1926-5
135. G. Santinello, C.W.T. Blackwell and Ph. Weller (eds.): *Models of the History of Philosophy.* Vol. 1: From its Origins in the Renaissance to the 'Historia Philosophica'. 1993 ISBN 0-7923-2200-2
136. M.J. Petry (ed.): *Hegel and Newtonianism.* 1993 ISBN 0-7923-2202-9

ARCHIVES INTERNATIONALES D'HISTOIRE DES IDÉES
*
INTERNATIONAL ARCHIVES OF THE HISTORY OF IDEAS

137. Otto von Guericke: *The New (so-called Magdeburg) Experiments* [Experimenta Nova, Amsterdam 1672]. Translated and edited by M.G. Foley Ames. 1994 ISBN 0-7923-2399-8
138. R.H. Popkin and G.M. Weiner (eds.): *Jewish Christians and Cristian Jews.* From the Renaissance to the Enlightenment. 1994 ISBN 0-7923-2452-8
139. J.E. Force and R.H. Popkin (eds.): *The Books of Nature and Scripture.* Recent Essays on Natural Philosophy, Theology, and Biblical Criticism in the Netherlands of Spinoza's Time and the British Isles of Newton's Time. 1994 ISBN 0-7923-2467-6
140. P. Rattansi and A. Clericuzio (eds.): *Alchemy and Chemistry in the 16th and 17th Centuries.* 1994 ISBN 0-7923-2573-7
141. S. Jayne: *Plato in Renaissance England.* 1995 ISBN 0-7923-3060-9
142. A.P. Coudert: *Leibniz and the Kabbalah.* 1995 ISBN 0-7923-3114-1
143. M.H. Hoffheimer: *Eduard Gans and the Hegelian Philosophy of Law.* 1995 ISBN 0-7923-3114-1
144. J.R.M. Neto: *The Christianization of Pyrrhonism.* Scepticism and Faith in Pascal, Kierkegaard, and Shestov. 1995 ISBN 0-7923-3381-0
145. R.H. Popkin (ed.): *Scepticism in the History of Philosophy.* A Pan-American Dialogue. 1996 ISBN 0-7923-3769-7
146. M. de Baar, M. Löwensteyn, M. Monteiro and A.A. Sneller (eds.): *Choosing the Better Part.* Anna Maria van Schurman (1607–1678). 1995 ISBN 0-7923-3799-9
147. M. Degenaar: *Molyneux's Problem.* Three Centuries of Discussion on the Perception of Forms. 1996 ISBN 0-7923-3934-7
148. S. Berti, F. Charles-Daubert and R.H. Popkin (eds.): *Heterodoxy, Spinozism, and Free Thought in Early-Eighteenth-Century Europe.* Studies on the *Traité des trois imposteurs.* 1996 ISBN 0-7923-4192-9
149. G.K. Browning (ed.): *Hegel's* Phenomenology of Spirit: *A Reappraisal.* 1997 ISBN 0-7923-4480-4
150. G.A.J. Rogers, J.M. Vienne and Y.C. Zarka (eds.): *The Cambridge Platonists in Philosophical Context.* Politics, Metaphysics and Religion. 1997 ISBN 0-7923-4530-4
151. R.L. Williams: *The Letters of Dominique Chaix, Botanist-Curé.* 1997 ISBN 0-7923-4615-7
152. R.H. Popkin, E. de Olaso and G. Tonelli (eds.): *Scepticism in the Enlightenment.* 1997 ISBN 0-7923-4643-2
153. L. de la Forge. Translated and edited by D.M. Clarke: *Treatise on the Human Mind (1664).* 1997 ISBN 0-7923-4778-1
154. S.P. Foster: *Melancholy Duty.* The Hume-Gibbon Attack on Christianity. 1997 ISBN 0-7923-4785-4
155. J. van der Zande and R.H. Popkin (eds.): *The Skeptical Tradition Around 1800.* Skepticism in Philosophy, Science, and Society. 1997 ISBN 0-7923-4846-X
156. P. Ferretti: *A Russian Advocate of Peace: Vasilii Malinovskii (1765–1814).* 1997 ISBN 0-7923-4846-6
157. M. Goldish: *Judaism in the Theology of Sir Isaac Newton.* 1998 ISBN 0-7923-4996-2
158. A.P. Coudert, R.H. Popkin and G.M. Weiner (eds.): *Leibniz, Mysticism and Religion.* 1998 ISBN 0-7923-5223-8
159. B. Fridén: *Rousseau's Economic Philosophy.* Beyond the Market of Innocents. 1998 ISBN 0-7923-5270-X
160. C.F. Fowler O.P.: *Descartes on the Human Soul.* Philosophy and the Demands of Christian Doctrine. 1999 ISBN 0-7923-5473-7

ARCHIVES INTERNATIONALES D'HISTOIRE DES IDÉES
*
INTERNATIONAL ARCHIVES OF THE HISTORY OF IDEAS

161. J.E. Force and R.H. Popkin (eds.): *Newton and Religion*. Context, Nature and Influence. 1999 ISBN 0-7923-5744-2
162. J.V. Andreae: *Christianapolis*. Introduced and translated by E.H. Thompson. 1999 ISBN 0-7923-5745-0
163. A.P. Coudert, S. Hutton, R.H. Popkin and G.M. Weiner (eds.): *Judaeo-Christian Intellectual Culture in the Seventeenth Century*. A Celebration of the Library of Narcissus Marsh (1638–1713). 1999 ISBN 0-7923-5789-2
164. T. Verbeek (ed.): *Johannes Clauberg* and Cartesian Philosophy in the Seventeenth Century. 1999 ISBN 0-7923-5831-7
165. A. Fix: *Fallen Angels*. Balthasar Bekker, Spirit Belief, and Confessionalism in the Seventeenth Century Dutch Republic. 1999 ISBN 0-7923-5876-7
166. S. Brown (ed.): *The Young Leibniz and his Philosophy (1646–76)*. 2000 ISBN 0-7923-5997-6
167. R. Ward: *The Life of Henry More*. Parts 1 and 2. 2000 ISBN 0-7923-6097-4
168. Z. Janowski: *Cartesian Theodicy*. Descartes' Quest for Certitude. 2000 ISBN 0-7923-6127-X
169. J.D. Popkin and R.H. Popkin (eds.): *The Abbé Grégoire and his World*. 2000 ISBN 0-7923-6247-0
170. C.G. Caffentzis: *Exciting the Industry of Mankind. George Berkeley's Philosophy of Money*. 2000 ISBN 0-7923-6297-7
171. A. Clericuzio: *Elements, Principles and Corpuscles*. A Study of Atomisms and Chemistry in the Seventeenth Century. 2001 ISBN 0-7923-6782-0
172. H. Hotson: *Paradise Postponed*. Johann Heinrich Alsted and the Birth of Calvinist Millenarianism. 2001 ISBN 0-7923-6787-1
173. M. Goldish and R.H. Popkin (eds.): *Millenarianism and Messianism in Early Modern European Culture*. Volume I. Jewish Messianism in the Early Modern World. 2001 ISBN 0-7923-6850-9
174. K.A. Kottman (ed.): *Millenarianism and Messianism in Early Modern European Culture*. Volume II. Catholic Millenarianism: From Savonarola to the Abbé Grégoire. 2001 ISBN 0-7923-6849-5
175. J.E. Force and R.H. Popkin (eds.): *Millenarianism and Messianism in Early Modern European Culture*. Volume III. The Millenarian Turn: Millenarian Contexts of Science, Politics and Everyday Anglo-American Life in the Seventeenth and Eighteenth Centuries. 2001 ISBN 0-7923-6848-7
176. J.C. Laursen and R.H. Popkin (eds.): *Millenarianism and Messianism in Early Modern European Culture*. Volume IV. Continental Millenarians: Protestants, Catholics, Heretics. 2001 ISBN 0-7923-6847-9
177. C. von Linné: *Nemesis Divina*. (edited and translated with explanatory notes by M.J. Petry). 2001 ISBN 0-7923-6820-7
178. M.A. Badía Cabrera: *Hume's Reflection on Religion*. 2001 ISBN 0-7923-7024-4
179. R.L. Williams: *Botanophilia in Eighteenth-Century France*. The Spirit of the Enlightenment. 2001 ISBN 0-7923-6886-X
180. R. Crocker (ed.): *Religion, Reason and Nature in Early Modern Europe*. 2001 ISBN 1-4020-0047-2

ARCHIVES INTERNATIONALES D'HISTOIRE DES IDÉES
*
INTERNATIONAL ARCHIVES OF THE HISTORY OF IDEAS

181. E. Slowik: *Cartesian Spacetime*. Descartes' Physics and the Relational Theory of Space and Motion. 2002 ISBN 1-4020-0265-3

KLUWER ACADEMIC PUBLISHERS – DORDRECHT / BOSTON / LONDON